This book is to be returned on or be
th. .ate sta ped b w.

NC

DIVERSITY OF ORGANISMS

The S203 Course Team

DIVERSITY OF ORGANISMS

Edited by Caroline M. Pond

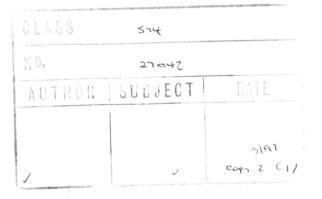
BIOLOGY: FORM AND FUNCTION

Hodder & Stoughton The Open University

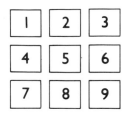

Cover illustrations

1 Polar bear, *Ursus maritimus* (phylum Chordata, class Mammalia, order Carnivora, family Ursidae) from around Hudson Bay, Canada.

2 Thorns of ant acacia, *Acacia depranolobium* (division Chlorophyta, subclass Angiospermidae, family Leguminosae) from savannah grasslands of Kenya.

3 Adult cricket (phylum Arthropoda, class Insecta, order Orthoptera) from Ugandan forests.

4 *Euglena* (kingdom Plantae, division Euglenophyta *or* kingdom Protista, class Mastigophora, order Euglenoidida) in freshwater.

5 Spider (phylum Arthropoda, class Arachnida, order Araneae) from rainforests of central America.

6 Bugs (phylum Arthropoda, class Insecta, order Hemiptera, family Coreidae) from rainforests of central America.

7 Red–blue–green macaw, *Ara chloroptera* (phylum Chordata, class Aves, order Psittaciformes) from rainforests of Central and South America.

8 Brown seaweed, *Laminaria* (kingdom Plantae, division Phaeophyta) on the lower shores of North Atlantic coasts.

9 Praying mantis (phylum Arthropoda, class Insecta, order Dictyoptera, family Mantidae) from West African rainforests.

Back cover: *Heliconius* butterfly (phylum Arthropoda, class Insecta, order Lepidoptera) from Central America.

British Library Cataloguing in Publication Data
Diversity of organisms.
1. Biology
I. Pond, Caroline M. II. Series
574

ISBN 0–340–53189–4

First published 1990.

Designed by the Graphic Design Group of the Open University.

The text forms part of an Open University course. Further information on Open University courses may be obtained from the Admissions Office, The Open University, P.O. Box 48, Walton Hall, Milton Keynes, MK7 6AB.

Typeset by Wearside Tradespools, Fulwell, Sunderland, printed in Great Britain by Thomson Litho Ltd, East Kilbride for the educational division of Hodder and Stoughton Ltd, Mill Road, Dunton Green, Sevenoaks, Kent TN13 2YA, in association with the Open University, Walton Hall, Milton Keynes, MK7 6AB.

CONTENTS

PREFACE

Diversity of Organisms is the first in a series of five volumes that provide a general introduction to biology. It is designed so that it can be read on its own (like any other textbook) or studied as part of *S203, Biology: Form and Function*, a second level course for Open University students. As well as the five books, the course consists of five associated study texts, 30 television programmes, several audiocassettes and a series of home experiments. As is the case with other Open University courses, students of S203 are required both to complete written assignments during the year and to sit an examination at the end of the course.

In this book, each subject is introduced in a way that makes it readily accessible to readers without any specific knowledge of that area. The major learning objectives are listed at the end of each chapter, and there are questions (with answers given at the end of the book) which allow readers to assess how well they have achieved these objectives. Key words are identified in bold type both in the text where they are explained and also in the index, for ease of reference. A 'further reading' list is included for those who wish to pursue certain topics beyond the limits of this book.

INTRODUCTION

Biology is the study of the origins, mechanisms and interactions of living organisms. Such biological processes cannot be deduced directly from a mathematical, physical or philosophical system: they are based upon observations and experimentation on natural living systems. Theories and generalizations are as important in biology as they are in any other science, but biological theories must be based upon specific examples. Scientists cannot study biological processes or propose or assess biological theories without an extensive knowledge of the structure and function of a wide range of organisms.

There are several millions of different kinds of organisms, most of them unfamiliar to non-specialists. Many kinds of organisms do not have common names, or the names are applied indiscriminately to several different kinds of organisms. Thus to many people the term 'insect' means any small terrestrial creature with several legs, and is applied to spiders, ticks, mites and woodlice (which biologists believe are fundamentally different from true insects) but not to some genuine insects (e.g. maggots) that lack legs and so are often mistaken for worms. An unambiguous and internationally recognized system of naming all organisms is therefore an essential foundation for all biological studies. The science of naming and classifying organisms is called taxonomy.

Naming organisms

A species is defined as a group of organisms that interbreed to produce fertile offspring. However, in the great majority of cases, this stringent criterion cannot be applied and, in practice, the identification and description of species is based upon anatomical characters. In the eighteenth century, the Swedish biologist Linnaeus developed a system for naming organisms in which each species is given a **binomial** name. The first name is that of the genus (plural: genera) and is always written with a capital letter. The second name is that of the particular species within the genus and never has a capital letter, even when it is derived from the name of a person or place. The generic name may be shortened to an initial if it is clear which genus is meant, and, if the species name is unknown, the organism may be identified by the generic name and 'sp.'. For example, *Limax* sp. indicates one of several very similar species of slug (Plate 13d). Binomial species names are always written in italics (or underlined in handwriting).

Some genera include only one species (e.g. the African grey parrot, *Psittacus erithacus* (Plate 17c) and the giant panda, *Ailuropoda melanoleuca*) while others such as *Aphis* (aphids) and *Bruchis* (weevils) include hundreds of species. As far as possible, generic names are unique to a particular genus, but specific names such as *vulgaris* (common), *maximus* (largest), *repens* (creeping) occur in the scientific names of many different organisms. The ending of the species name may be modified so that it 'agrees' with the generic name (e.g. *Sturnus vulgaris*, the starling, but *Patella vulgata*, the common limpet) but binomial names are never altered to form plurals.

to their degree of similarity. Until recently, taxonomists worked mainly with dead, often pickled specimens, so the main kind of information at their disposal was gross anatomy and microscopic structure, so anatomical features were the chief basis for both natural and cladistic classification.

The lack of information about how and when new taxonomic categories appear means that all schemes incorporate errors and so they are constantly being revised. Furthermore, experts may hold different opinions on points of fact and of interpretation, and hence may favour different schemes. You will probably come across discrepancies between the classification scheme used here, and those in other textbooks. Since the exact definition of all groups is to some extent a matter of opinion, species are reshuffled into genera and between the larger categories as new information becomes available, and ideas change about the relative importance of characters. For example, in 1774, a taxonomist called Phipps classified polar bears in the genus *Ursus*, which includes most other bears, and gave them the specific name *U. maritimus*. In 1962, taxonomists who were impressed by the contrasts in diet and habits between polar bears and other bears, proposed reclassifying the arctic species in a separate genus, naming it *Thalarctos* ('sea bear') *maritimus*. However, when polar bears and brown bears kept together in zoos were observed to mate and produce viable offspring, and analysis of proteins in the blood revealed close genetic similarities between the two species, it became clear that classifying polar bears and other bears in separate genera did not reflect the true biological relationship between the species. So, in the 1970s, polar bears were reinstated in the genus *Ursus* and are now known by their original name, *Ursus maritimus* Phipps. When reclassification in a different genus is adopted, the taxonomist's name is written in brackets; thus the scientific name of the brown trout is *Salmo trutta* L., but the speckled trout, which resembles the char more closely than other kinds of trout, was reclassified in the genus *Salvelinus* and is therefore called *Salvelinus fontinalis* (Mitchill).

The principal 'higher' taxonomic categories are: **kingdom**, **phylum** (in plant taxonomy, the term '**division**' replaces 'phylum'), **class**, **order** and **family**. The prefixes 'sub' and 'super' are sometimes applied to subdivisions and groupings of these categories. The names of higher categories are often synthesized from Latin or Greek words in much the same way as generic or species names are constructed. The Latinized versions of such names normally end in '-a', '-ae' or '-es' and always start with a capital letter. The English versions of the same words usually end in -s or -an and never have a capital letter except at the beginning of a sentence. Thus the Latinized form 'Prokaryota' becomes 'prokaryotes' in English and 'Amphibia' becomes 'amphibian'. Taxonomic terms above the level of genus and species are never written in italics. The easiest way to understand the kinds of criteria that are used to define these 'higher' taxonomic categories is to study some actual examples, such as those described in Chapter 1.

As well as providing an introduction to the general biology of micro-organisms, plants and animals, this book tries to explain the origins of biological diversity and the practical importance of a thorough knowledge of the diversity of organisms.

Chapter 1 is an introduction to the structure, life cycles and habits of the major groups of organisms, with special reference to the features used in their classification. We cut across these taxonomic categories in Chapter 2 to discuss some of the methods and concepts used for interpreting observations and designing experiments on living organisms. The importance of integrating laboratory and field studies is emphasized. Some of the biological factors that have promoted and maintained such a huge variety of animals, plants and micro-organisms are also explained, using examples from recent research.

Chapter 3 discusses some practical applications of knowledge about a wide range of organisms and illustrates why a knowledge of the diversity of organisms, as well as of basic physiological mechanisms, is an essential part of biology. Case histories taken from original research literature are used to show how the comparative method is used to elucidate biological problems that cannot be solved by experimentation on a single species.

The Appendix, A Survey of Living Organisms, is intended to be a supplement to the other chapters and should be used mainly for reference. The classification in the Appendix is much more complex and detailed than that used in the rest of the book.

as the tissues of more organisms are examined thoroughly; viruses have now been found in the cells of a wide range of wild vertebrates including oceanic fishes, and in insects, many plants, fungi, algae and bacteria.

Many viruses live in more than one species but do not necessarily have similar effects on the well-being of all their hosts. The virus that causes rabies has been found in a wide variety of mammals. Infected dogs become 'mad' and develop symptoms similar to the deadly human disease, but some other mammals, notably bats, are apparently unaffected even by high concentrations of the virus. Bats are therefore a major natural reservoir of the rabies virus. Another important group of viruses multiplies in both humans and in blood-sucking insects (e.g. fleas, mosquitoes, bedbugs) or ticks. Many such viruses are pathogenic, so the insects or ticks act as vectors, transmitting the disease from person to person.

Many plant diseases are caused by viruses, including those affecting such economically important species as strawberries, tomatoes, potatoes and squashes. Many such viruses are transmitted by herbivorous insects in much the same way as blood-sucking insects act as vectors for mammalian viral diseases. The tobacco mosaic virus (TMV) causes brown blotches on the leaves of tobacco plants. In 1935, TMV (Figure 1.1j) became the first virus to be purified and crystallized. It has been studied intensively since that time because infected plants accumulate large quantities of virus—up to 80% of the total protein in the sap can be of viral origin—thereby providing biologists with an abundant and convenient source of readily purifiable virus. Unfortunately, the viruses of animal cells very rarely accumulate in such high concentrations, so biologists need large quantities of the host tissue, and have to follow a lengthy extraction procedure to obtain sufficient viral material for biochemical studies.

Because viruses are variable and are difficult to identify except as a pathogen or as a foreign body in a cell, they are not classified into species in the same way as other organisms (described in the Introduction), and so they do not have binomial names. They are often named after the person or place involved in their identification (e.g. Sendai virus; Sendai is a city in Japan) or for the symptoms or diseases that they cause (e.g. TMV, polio virus, HIV). There have recently been attempts to classify viruses into families and other higher categories, but such schemes have not been widely adopted.

1.2 THE CLASSIFICATION OF MICRO-ORGANISMS

In older classifications, all larger organisms that lacked the essential features of animals were classified as plants and those invisible to the naked eye were lumped together as '**micro-organisms**'. The invention of the electron microscope enabled biologists to study minute organisms in much greater detail and the new information brought radical changes in our understanding of their structure and physiology. These advances led to revision of the theories about how the various kinds of micro-organisms might have evolved and to proposals for several different classification schemes. In one such scheme, organisms are divided into two groups on the basis of the structure of their cells, the **Prokaryota** ('before nucleus') and the **Eukaryota** ('true nucleus').

The Dutch biologist van Leeuwenhoek described some exceptionally large prokaryotes (from his own mouth) in the late seventeenth century, but they were not studied thoroughly until improvements in microscope design in the mid-nineteenth century produced clear images magnified up to 1 000 times.

Figures 1.2 and 1.3 show the essential features of typical prokaryotes as revealed by the electron microscope, and Plate 1 (left) is a light micrograph of an exceptionally large cyanobacterium. Like all other kinds of cell, the prokaryotes have an outer membrane and the internal structure is not homogeneous. However, the bacterium and the cyanobacterium lack the membrane-bound nucleus and mitochondria that are present in eukaryotic cells. Prokaryotes are also normally much smaller than most eukaryotes, typically about 1 μm in diameter. These contrasts, particularly the presence or absence of nuclei and membrane-bound organelles in the cell (e.g. mitochondria or chloroplasts), are the basis for distinguishing prokaryotes from eukaryotes.

Prokaryotes are the smallest and simplest organisms that can complete their life cycles independently. They are also the oldest form of life still in existence; their fossil remains can be detected in rocks formed at least 3×10^9 years ago, long before there are any signs of eukaryotic cells. They are abundant and diverse today, and so far as we can tell have been a major component of ecosystems throughout evolutionary history. They are important to humans not only as agents of disease and decay, but also because manipulation of their remarkably flexible metabolism is the basis of many forms of biotechnology (see Chapter 3, Section 3.4).

Prokaryotes are sufficiently consistent in form and properties to be classified into species in the same way as higher organisms, and are known by binomial names (see Introduction). However, many prokaryotes can acquire or lose structural features and biochemical properties quite readily, so many similarities between organisms may not actually arise from common ancestry and hence the classification cannot be 'natural'. Nonetheless, there are many thousands of living species of prokaryotes including typical bacteria ('little rods'), many, but by no means all, of which are elongated or oval in shape, and the larger, more plant-like cyanobacteria ('blue bacteria'), also known as blue–green bacteria or blue–green algae.

However, during the 1970s detailed study of an obscure group of small organisms revealed fundamental differences between them and all other prokaryotes. Their ribosomal RNAs also have unique sequences, quite unlike

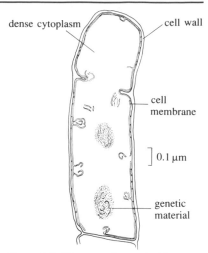

Figure 1.2 Diagram of a typical bacterium as seen with the electron microscope. One cell in a chain of similar cells is shown.

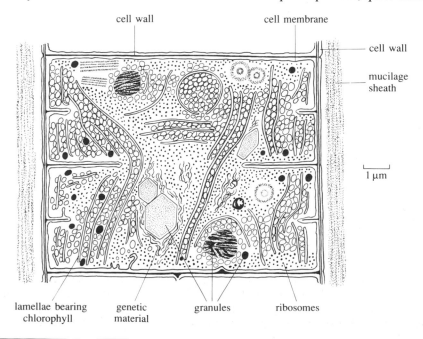

Figure 1.3 Diagram of a typical cyanobacterium, as seen with an electron microscope. One cell in a chain of similar cells is shown. Note the variety of granules in the cytoplasm.

that of any other known organisms, and there are fundamental differences in the structure of their cell walls, the outer cell membrane and the basic mechanism of transcription and translation of their genetic material. On the basis of this information, biologists have recently proposed dividing the prokaryotes into two kingdoms, the Eubacteria ('true bacteria') which include the bacteria (Section 1.2.1) and the cyanobacteria (Section 1.2.2) and the Archaebacteria ('ancient bacteria'). All archaebacteria are anaerobic and most live in 'extreme' habitats such as hot sulphurous springs, stagnant swamps and salty places like the Dead Sea, where few other kinds of organisms are able to survive. Some authorities believe that archaebacteria are a very ancient group, perhaps similar to some of the earliest organisms to evolve on Earth.

1.2.1 Bacteria

The internal structure of bacteria is relatively simple. Nutrients and respiratory gases are taken up across the thin, selectively permeable cell membrane (Figure 1.2), and waste materials and secretions pass out by the same route. There are no membrane-bound organelles such as mitochondria but there are numerous ribosomes which are often concentrated towards the middle of the cell. Many forms have one or more actively beating **flagella** (sing. flagellum, 'whip') with which they swim through fluids. The anatomical simplicity of their internal structure is deceptive—bacteria have some of the most varied and adaptable metabolic pathways known.

There are very few environments that do not support at least one species of bacterium. Some bacteria are **autotrophic** ('self feeding'). Like green plants, such organisms photosynthesize, that is, they use the energy of sunlight to synthesize complex organic compounds from atmospheric gases and other inorganic molecules that they absorb from their surroundings. Such bacteria, however, have an unusual form of photosynthesis that involves a light-absorbing pigment called bacterio-chlorophyll that differs in some fundamental ways from the chlorophyll of cyanobacteria (Section 1.2.2) and that of eukaryotic plants (Section 1.4). Other autotrophic bacteria obtain energy from the oxidation of simple inorganic compounds in the absence of light. Bacterial photosynthesis and many other autotrophic processes are anaerobic, and indeed, oxygen is lethally toxic to many such bacteria. Autotrophic bacteria require few, if any, complex organic molecules and a few kinds can obtain metabolic energy from compounds such as hydrogen sulphide that cannot be utilized by any other kind of organism. Such bacteria are therefore among the few kinds of organism to live permanently near volcanic craters and hot vents on the ocean floor and in other habitats that are as inhospitable as Hell to higher animals and plants.

Many other bacteria are **heterotrophic** ('other feeding'): they obtain both nutrients and energy by taking in organic molecules, usually derived from other organisms, and break them down to smaller components. Their nutrition is thus more like that of fungi and typical animals (see Sections 1.5 and 1.6). The majority of heterotrophic bacteria feed on excrement and on the bodies of other organisms after they have died. In fact, bacteria play a major role in the putrefaction and decomposition of animals and plants in almost all environments on land, sea and freshwater. Some species have very precise nutritional requirements, and their proliferation is easily curtailed by the lack of an essential nutrient, but most others can absorb and metabolize a wide range of substances. Some of the most widespread species can switch between alternative biochemical pathways within a few minutes.

As well as utilizing unusual nutrients, many bacteria synthesize and secrete organic compounds that are rare or unknown in plants or animals. Some such compounds are unpleasant or toxic to other organisms and probably serve to reduce competition for food between the bacterium and other organisms, particularly fungi and animals, by rendering the food unpalatable to them. Humans can smell or taste the presence of many of the bacteria that decompose fruit, leaves and the flesh of vertebrates. But others, such as *Clostridium botulinum* (which causes botulism), produce a toxin that is not readily detected by humans and are therefore particularly dangerous as a contaminant of food.

Many bacteria live in, on or near the tissues of other organisms. Some species live in animal guts (see Chapter 2, Section 2.4.1), such as the stomachs of cattle, sheep and deer and the hindguts of many insects and vertebrates, including humans, where they contribute to their hosts' digestion. One of the most thoroughly studied of all organisms is *Escherichia coli* (usually known as *E. coli*) which occurs naturally in the human colon (hindgut). Up to half the volume of human faeces consists of *E. coli* and other kinds of bacteria, or their dead remains, that have been living harmlessly inside the guts.

Asexual reproduction, in which the bacterium simply divides equally and each daughter cell grows to the 'adult' size before dividing again, can take place as frequently as three times an hour. Sometimes a simple form of sexual reproduction takes place; two bacteria join together and exchange copies of all or part of their genetic material before dividing. In many species of bacteria, the daughter cells stick together to form clumps or ribbons of cells. Under favourable conditions, and while in contact with a suitable food source, bacteria can multiply very rapidly.

◇ Assuming a generation time of twenty minutes and that there is unlimited suitable food, how long would it take to form a ribbon 100 m long of bacteria each of which is 5 μm in length?

◆ Eight hours. A ribbon 100 m long contains 20 million bacteria, which would be produced from a single bacterium in 24 generations if the cells divided once every twenty minutes, and none dies.

Clearly, bacteria quickly become very numerous under suitable conditions. However, their abundance is normally limited by the availability of food and by predation from the many higher organisms that eat them. When food is insufficient or environmental conditions are unsuitable, many bacteria form spores and can remain dormant but protected from desiccation for many months or years until they resume feeding and multiplying.

Many bacteria have one or more flagella, that propel them through liquids at speeds of up to $1 \, \text{mm h}^{-1}$. Although bacterial flagella are similar in general form and function to the flagella of eukaryotic cells, their internal structure and mechanism of movement are unique to prokaryotes. Passive transport by wind, water movements and on or in the bodies of larger organisms, is more important for colonization of new sources of food.

Bacteria are familiar to most people as agents of disease, including tonsilitis, diphtheria, typhoid, cholera, tuberculosis, some forms of pneumonia, gangrene, abscesses and boils, but only a small minority of bacteria are harmful to the organisms in which they live. Such pathogenic bacteria usually have more elaborate and specific nutrient requirements and less impressive powers of biochemical synthesis than free-living or symbiotic bacteria but, if provided with suitable artificial food, they can be cultured in the laboratory. Many kinds also survive in other species of animals, including laboratory rats and mice, in which their presence often, but not invariably, causes symptoms of disease similar to those observed in humans (see Chapter 3, Section 3.3.3).

As for most biochemical systems, the rate of growth of bacterial colonies depends upon the temperature, but the range of temperatures that they can tolerate varies widely between species. Some pathogenic bacteria flourish only at normal mammalian body temperature, while many free-living forms can survive prolonged exposure to high temperature, freezing and desiccation. Nonetheless, the growth of many pathogenic bacteria is significantly curtailed by quite small rises in temperature and the great majority are destroyed by heating to the boiling point of water. Birds and mammals respond to the invasion of foreign bacteria by generating more heat and raising the body temperature, and infected reptiles seek out warm places and bright sunlight. The fever so generated makes the host animal feel very ill and can damage, sometimes irreversibly, delicate organs such as the brain, the kidneys and eye, but it does curtail the proliferation of the foreign bacteria.

The biochemical machinery of some pathogenic bacteria, for example those which cause typhus fever in humans and psittacosis in many wild birds including parrots, is greatly reduced and simplified. The most thoroughly studied of these highly specialized bacteria are called rickettsiae (after the American biologist, T. H. Ricketts, who died of typhus while studying its cause); they are exceptionally small bacteria, hardly bigger than large viruses, and do not replicate except inside the cells of their host. At the other end of the scale of complexity are the actinomycetes (Figure A.1* and Chapter 3, Section 3.4.2) which resemble eukaryotic fungi (Section 1.5) in both structure and habits. They are abundant in soil in association with the roots of flowering plants and in rotting wood, where they form branching colonies of cells and can survive for some time as dormant spores.

1.2.2 Cyanobacteria (blue–green bacteria or blue–green algae)

Cyanobacteria (Figure 1.3 and Plate 1), sometimes called blue–green bacteria or blue–green **algae**†, are a group of bacteria that have ecological and physiological similarities to green plants. Like eukaryotic algae, multicellular green plants and some other kinds of bacteria, they are autotrophs, harnessing the energy from sunlight to drive biosynthetic processes by means of photosynthesis. They resemble eukaryotic algae and terrestrial green plants but differ from other photosynthetic bacteria in containing **chlorophyll**. They synthesize complex organic compounds from carbon dioxide and water in sunlight, and, with adequate supplies of inorganic nutrients and sufficient light, can build all their body components. Like plants, but in contrast to other photosynthetic bacteria, cyanobacteria both release oxygen as a product of photosynthesis and use it in breaking down complex molecules.

Another way in which cyanobacteria are similar to plants is in having a multilayered cell wall and sometimes mucilage around the cell membrane. However, unlike higher plants, many cyanobacteria can also synthesize amino acids and other nitrogen-containing compounds from gaseous nitrogen

*Figures A.1–A.46 are in the Appendix to this book.

†The term 'alga' was used by early taxonomists to describe all the simple, free-living, densely coloured plant-like organisms. However, more thorough studies of the internal structure of these and other minute organisms has shown that this term describes at least three quite different kinds of organism which are more closely related to non-algal-like organisms than to each other. 'Alga' therefore describes a life form, like 'tree' or 'bush', rather than a taxonomic group.

absorbed from the atmosphere. As in other bacteria, nitrogen fixation is anaerobic. This ability to 'fix' nitrogen is found in other groups of prokaryotes but never in eukaryotes. Most cyanobacteria are about $0.2-5.0\,\mu m$ in diameter, and are thus generally larger than other bacteria, although of course much smaller than most eukaryotic cells. The cells reproduce by division but sexual reproduction has never been observed. Cyanobacteria do not have flagella but some forms can move actively by gliding over a damp surface. They often contain other pigments besides chlorophyll, so, in spite of their alternative name, 'blue–green', cyanobacteria can be any colour from blackish green to orange.

Cyanobacteria are less familiar to most people than are other prokaryotes because their ecological relationships with humans and domestic livestock are more remote than those of heterotrophic bacteria. Nonetheless, they occur in a wide variety of places exposed to sunlight, such as in the upper waters of lakes and the oceans and as encrustations on rocks, plants and sedentary animals. Although the individual cells are invisible to the naked eye, cyanobacteria often form ribbons, sheets or clumps (Figure A.2). Cyanobacteria occur in the sea and in freshwater where they float in the surface waters or live attached to rocks or other organisms. Under suitable conditions, such as stagnant, nutrient-rich ponds and lakes, they become very abundant. The green scum that grows on the shores of lakes and reservoirs, and on the sides of aquaria exposed to sunlight, is mostly cyanobacteria. In bright light they produce oxygen from photosynthesis but in darkness, and when present in large numbers, they deplete the dissolved oxygen, rendering the water inhospitable to other plants and animals. They also occur on rocks and mud on sea-shores and in damp habitats on land, including soil, rock crevices, and in and around thermal springs where the high temperatures and high concentrations of toxic gases exclude most eukaryotic plants. Their outer covering of mucilage often makes surfaces covered with cyanobacteria extremely slippery to walk on.

Organisms that live free in the upper waters but that are unable to swim actively against ocean currents are called **plankton**. Most such organisms are small, but a few, such as certain jellyfishes (see Section 1.6.2) are very large. The cyanobacteria are very important primary producers in the plankton of oceans and lakes; they are eaten by a variety of larger planktonic organisms, such as protistans (Section 1.3.1), which are in turn eaten by many other marine animals, including large fishes and whales. Light is absorbed even by pure, clean water, so aquatic cyanobacteria are most abundant in the top 2 m of water, but some species can photosynthesize efficiently at low light intensity and so occur in deeper water. Many can form gas-filled vesicles that may enable them to float on or near the surface. Some planktonic cyanobacteria live as single cells (Plate 1), but many others form filaments.

Summary of Sections 1.1 and 1.2

Viruses consist only of protein and genetic material. They only replicate inside bacterial, plant or animal cells.

Prokaryotes are the smallest organisms that can complete their life cycle independently. They lack a nucleus and membrane-bound organelles but they have diverse biochemical capacities. Bacteria may be autotrophic or heterotrophic and occur in a huge range of environments, including places where few other organisms can survive, and are abundant on the dead remains of plants and animals. A minority live in or on other living organisms where they may cause disease. Cyanobacteria are non-motile and photosynthesize like higher plants and occur in many places that are moist and exposed to sunlight. Archaebacteria differ fundamentally from other prokaryotes and are mainly autotrophs, living in extreme habitats.

Question 1 (*Objective 1.2*) Which of the following are present in viruses? Cell wall, membranes, nucleus, mitochondria, proteins, DNA or RNA, glucose, pigments.

Question 2 (*Objective 1.3*) Complete Table 1.1, putting a + if the feature is normally present in the group, and a − if it is absent.

Table 1.1 A comparison of prokaryotes and eukaryotes (for use with Question 2).

	Prokaryotes	Eukaryotes
An outer membrane around the cell		
Genetic material mainly confined to a membrane-bound nucleus		
Organelles such as chloroplasts and mitochondria		
Enzymes for aerobic respiration		
Capacity for exchanging genetic material (sexual reproduction)		
Capacity for cell division by mitosis		

Question 3 (*Objective 1.4*) Which of the features (i)–(vi) are shown by the following? (a) all prokaryotes; (b) some bacteria and all cyanobacteria; (c) some bacteria and some cyanobacteria; (d) some bacteria but never by cyanobacteria; (e) cyanobacteria only; (f) neither bacteria nor cyanobacteria.

   (i) The ability to synthesize complex organic molecules from carbon dioxide by photosynthesis
  (ii) The ability to fix nitrogen
 (iii) The ability to move by beating flagella
 (iv) The possession of chlorophyll similar to that of higher plants
  (v) Division into two similar daughter cells
 (vi) Exchange of genetic material (sexual reproduction)
(vii) Internal cytoplasmic membranes

1.3 EUKARYOTES

Fossils of eukaryotic cells are found in rocks laid down in shallow seas at least 1.4×10^9 years ago. The principal differences between eukaryotic and pro-karyotic cells are in size and the fact that the former contain a nucleus and clearly differentiated organelles bound by internal membranes. Most eu-karyotic cells are about a thousand times larger than typical prokaryotic cells, usually about 10^{-15}–10^{-8} litres in volume (diameter 1–100 μm), but a few highly specialized cells, such as birds' eggs, are up to 1 litre in volume (i.e. ostrich eggs). The most widespread organelles are the nucleus, mitochondria and chloroplasts, but these structures are not found in all eukaryotic cells. Ribosomes are also usually associated with internal membranes. The genetic material is always DNA, most of which is located in the nucleus. Eukaryotic cells divide by mitosis: the daughter cells each receive a copy of all the genetic material and about half the cytoplasm of the parent cell. Genetic material is exchanged only following a more elaborate form of cell division, **meiosis**, which results in the formation of gametes (which are haploid) that subsequently fuse (in plants, the products of meiosis form haploid organisms which then produce gametes by mitosis, see Section 1.4).

Most **unicellular** eukaryotes exist as single cells throughout their life cycle but, in some species, groups or strings of similar cells remain attached to each other, forming a **colony**. If a unicellular colony is divided, all the fragments are normally able to survive and complete their life cycle. Other eukaryotes, including all terrestrial plants and large animals, are **multicellular**; the cells of multicellular organisms are not only in mechanical contact with each other, individual cells or groups of cells develop different and specialized structures and complementary functions. In most cases, the cells of a multicellular organism are tightly integrated together, isolated fragments of most multicellular organisms survive for only a short time and do not complete the life cycle. The classification of unicellular and simple multicellular eukaryotes is complicated. Several different schemes have been proposed, no one of which clearly stands out as easier to use or reflecting more accurately the evolutionary relationships between the organisms.

The discovery of archaebacteria has led to a major shift in opinion on how the living world should be classified into kingdoms. During the last twenty years, various revisions have been discussed, proposing from three to thirteen kingdoms. Cell biologists are satisfied with only two groups, Eukaryota and Prokaryota, based upon the structure of the cells rather than of the whole organisms, but ecologists and physiologists generally prefer six kingdoms: Archaebacteria, Eubacteria (Section 1.2), Protista, Plantae, Fungi, Animalia, and this scheme is adopted here. The definitions of these kingdoms may overlap with terms such as multicellular and unicellular as illustrated in Figure 1.4.

The unicellular eukaryotic organisms include the Protista ('very first things'), which resemble animals in internal structure, diet and habits, and several groups of plant-like unicellular algae. Organisms commonly known as 'plants' are divided into two kingdoms, the Fungi, which never photosynthesize, and the Plantae, which, except for a few parasitic species, are autotrophic and photosynthesize aerobically in sunlight. Although traditionally regarded as atypical plants and studied by botanists, recent research has shown that fungi are so different from animals and plants that this unique, and in many ways

Figure 1.4 Summary of the grades of organization of the major categories of organisms.

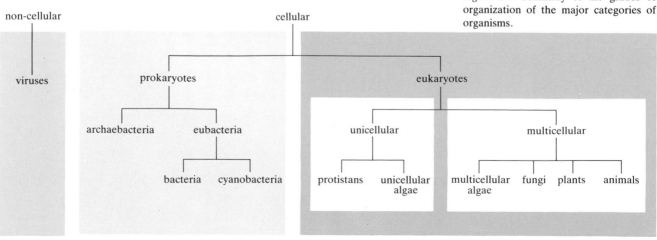

highly successful, group is now regarded as a separate kingdom (see Section 1.5). Botanists now recognize at least seven **divisions** in the kingdom Plantae, four of which include only unicellular algae. There are two divisions that include larger, multicellular algae, and all the terrestrial green plants and certain other algae are classified together in a single division, the Chlorophyta ('green plants') (see Section 1.4).

All multicellular, heterotrophic organisms (except the few parasitic plants) other than fungi are placed in the kingdom Animalia. Instead of divisions, zoologists recognize about 25 **phyla** (singular: phylum) in the animal kingdom, established according to somewhat different criteria from those used to define plant divisions. This number of basic categories may seem excessively large, but with about a million living species, and at least a million more extinct forms, animals are far more diverse than plants. Some factors that may have promoted the evolution of such a huge variety of different animals are discussed in Chapter 2. Some phyla include only a few, relatively obscure species; there is space for detailed treatment of only a few of the most abundant and familiar phyla in this chapter. The principal category within a phylum or division is the **class**; classes are divided into **orders** and orders into **families**. Families may include from one to many scores of **genera**. There are brief descriptions of all the divisions of the plant kingdom and of many of the phyla and classes of the animal kingdom in the Appendix.

1.3.1 Protistans

The unicellular protistans are some of the most abundant eukaryotic organisms. They occur in an immense variety of sizes, shapes and colours (Figures 1.5, A.15–17 and Plate 2) and are abundant as free-living organisms in the ocean, in freshwater, in soil and as parasites and mutualistic symbionts with animals (see Chapter 2). Most protistans range in size from 5 μm to 1 mm in diameter, but a few kinds can be much larger. Many forms have one or more flagella that contain contractile proteins, some of which resemble those in the muscles of animals, and differ in some important ways from prokaryotic flagella. Several groups of protistans, known collectively as **flagellates**, may have at least one flagellum at some stage of their life history. Relatively short, numerous flagella are called **cilia** ('eyelash') and protistans bearing them are called ciliates. Most ciliate and flagellate protistans can swim actively, some at speeds of several body lengths per second.

Many other protistans, such as *Amoeba* (Figure 1.5d) do not have cilia or flagella and move by means of contractile proteins in the cytoplasm that form temporary protrusions, **pseudopodia** ('false feet'; sing. pseudopodium). Many protistans, particularly marine species in the class Sarcodina, have elaborate calcareous or silica skeletons (e.g. Figure 1.5c). The form and composition of the skeleton are important criteria for the classification of these groups of protistans.

Protistans usually reproduce asexually by division of a mature cell into two 'daughter cells'. Many species can also reproduce sexually, with two cells exchanging genetic material before or during cell division. A distinctive feature of ciliates is that they have two sorts of nuclei, meganuclei and micronuclei (Figures 1.5a and b and A.17). Sexual reproduction has been most thoroughly studied in the ciliate *Paramecium* (Figure 1a and Plate 2a). In this species, partners join together at a particular part of the cell surface and exchange genetic material. Although they appear very similar to human observers, *Paramecium* occurs in at least five different 'mating types', and individuals mate only with members of a different mating type from their own.

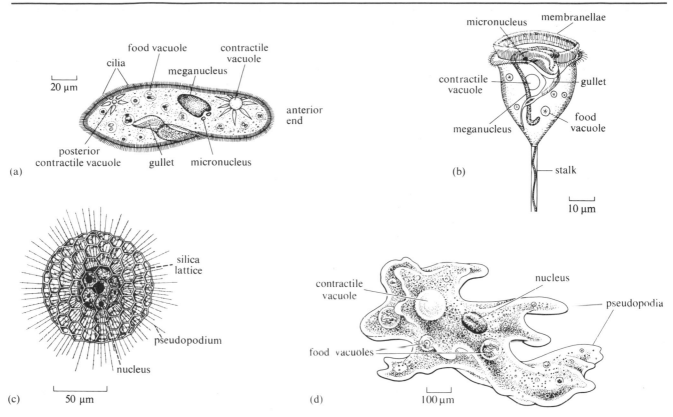

Figure 1.5 Various protistans. (a) *Paramecium* (Ciliata). (b) *Vorticella* (Ciliata). (c) *Heliosphaera* (Sarcodina, Radiolaria). (d) *Amoeba* (Sarcodina).

Almost all protistans feed on organic material but their diet ranges from absorbing small molecules from their surroundings to engulfing large particles, including other cells, by a process called **phagocytosis**: the cell membrane forms an invagination that encloses the food particle, and forms a cytoplasmic vacuole around it. Digestive enzymes are secreted into the vacuole, and the digested nutrients are absorbed. In many protistans, phagocytosis can occur anywhere on the cell surface, but in others, such as *Paramecium*, the process occurs only through one or more specialized sites. When suitable food is available, many protistans can feed voraciously and multiply rapidly. For example, each amoeba-like cell of *Dictyostelium* eats about a thousand bacteria every time it doubles in size and divides.

Protistans are very abundant in the soil, and in the upper waters of lakes, rivers and oceans, where they feed on bacteria, particularly cyanobacteria, and the dead remains of larger organisms. In most soils, flagellates are by far the most numerous kind of protistan, and ciliates are relatively rare, but representatives of all three groups are common in water. Figure 1.5 shows a few kinds of aquatic protistans. There is much variation in form and size, and some species contain an internal skeleton or are enclosed within a 'shell'. These hard parts support the living material and may aid in flotation and/or facilitate diffusion of nutrients from the water to the centre of the organism. Radiolarians (Figure 1.5c) are large sarcodine protistans that are abundant in the marine plankton, particularly in warm oceans. Much of the living material is surrounded by an intricate and beautiful skeleton, but there are also long thin pseudopods protruding through the shell. A few kinds of marine and freshwater protistans attach themselves to fixed objects such as rocks and seaweeds. *Vorticella* (Figure 1.5b and Plate 2b) is a freshwater ciliate that lives attached to rocks or vegetation. Its rings of cilia propel water currents towards a specialized region of the cell membrane through which food particles enter.

Protistans are also abundant in the guts of many vertebrates, including humans, and some insects, notably wood-boring termites, where they aid digestion by producing enzymes that their hosts cannot synthesize. Such symbiotic protistans are particularly important in the forestomachs of cattle and sheep (see Chapter 2, Section 2.4.2). Very few protistans are harmful to humans or domestic animals. However, the few human diseases caused by protistans (malaria, sleeping sickness and amoebic dysentery) are very widespread and difficult to eradicate or cure (see Chapter 3, Section 3.4.2).

Many protistans can survive long periods without food or suitable living conditions by becoming dormant and secreting materials that encase them in an impermeable **cyst**.

◇ In what kinds of habitats would you expect to find protistans that can form cysts?

◆ Protistans that live in places that are susceptible to drying, such as the soil, in freshwater and on the sea-shore, and parasitic species. Oceanic protistans would be less likely to form dormant cysts.

1.3.2 Unicellular algae

Although their simple structures appear to be similar, unicellular eukaryotic algae differ in the chemical composition of their chloroplast pigments, cell walls and food storage materials and are therefore placed in different divisions of the plant kingdom. The main kinds are the aquatic diatoms and dinoflagellates illustrated in Figure 1.6 and Plate 3, and a few terrestrial forms that are classified with other 'green algae' in the Division Chlorophyta, such as *Chlorella*, which forms 'scum' on tree bark and other moist surfaces. The formal classification is complicated, with aquatic algae being divided into several taxonomic groups.

◇ Referring to Section 1.3.1, and to Figures 1.5 and 1.6 and Plates 2 and 3, list some similarities and contrasts between these unicellular algae and protistans.

◆ Like protistans, some species have flagella, siliceous spicules or internal skeletons. Unlike protistans, these algae have one or more chloroplasts.

All algal chloroplasts contain chlorophyll, but many eukaryotic algae also contain other pigments which may mask the green pigment. Thus many algae are brown or golden-brown as well as green, and their pigments are the basis for their formal classification, as well as for many of their common names. Algae of different colours photosynthesize most efficiently at different light intensities and hence live at different depths and positions in the water. Like most organisms that do not feed or photosynthesize continuously, algae store energy-rich compounds in the form of particles; the chemical composition and arrangement in the cell of these storage particles differs between groups of algae and is also used as a basis for their classification.

Unicellular algae reproduce mainly by asexual mitotic division of the cells. In some unicellular algae, individual cells live singly in the plankton, but in many other species, for example, *Zygnema* (Plate 1) and the familiar freshwater

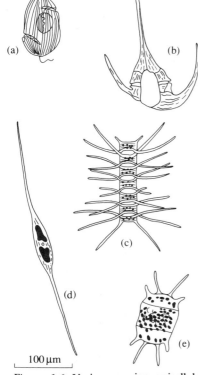

100 μm

Figure 1.6 Various marine unicellular algae. (a) and (b) Dinoflagellates. (c), (d) and (e) Diatoms. These species are relatively large. Many other unicellular algae are much smaller.

'weed' *Spirogyra* (Plate 3c), the cells remain together and form long filaments. *Chlamydomonas*, *Euglena* (Plate 3a) and *Volvox* (Plate 3d) are some of the most widely studied algae. They all have flagella and can swim actively. *Chlamydomonas* and *Euglena* occur as single cells, but clusters of similar *Volvox* cells remain together, forming a spherical colony. These three species live in freshwater but similar-looking forms live in the sea.

◇ Compare the structures of the algae *Euglena* and *Chlamydomonas* (Figures 1.7a and 1.7b) with those of the protistans *Peranema* and *Polytoma* (Figures 1.8a and 1.8b).

◆ These organisms are remarkably similar in size and structure. *Euglena* and *Chlamydomonas* are regarded as plants and *Peranema* and *Polytoma* as protistans only because the former have chloroplasts but the latter are incapable of photosynthesis.

Most unicellular algae occur in the upper waters of oceans and lakes but some live on submerged or temporarily exposed rocks, plants and the shells of dead and living animals. The film of algae and other micro-organisms growing on such structures is an important source of food for small grazing invertebrates. An example of this type of alga is *Acetabularia*, an exceptionally large species (up to 5 cm long) which has been used for research in cell biology.

◇ Summarize the reasoning behind classifying unicellular algae with cyano-bacteria, protistans or multicellular green plants. Why is the scheme in Figure 1.4 probably closest to the natural classification?

◆ The ability to photosynthesize and the fact that most live in well-lit places link all algae together. However, features such as active swimming with flagella (e.g. *Euglena*) and inorganic skeletons link some of them with protistans. The presence of chloroplasts in all unicellular algae and rigid cell walls in most forms show their similarity to the multicellular green plants. The scheme shown in Figure 1.4 gives high priority to cellular organization and to biochemical factors, such as photosynthetic pigments.

Planktonic unicellular algae multiply rapidly in well-lit waters in which there are sufficient supplies of inorganic nutrients, particularly phosphates, nitrates plus some calcium and iron. They are very abundant in both freshwater and the sea, and are the major food for protistans and many kinds of small multicellular animals (see Section 1.6). Photosynthesis by marine unicellular

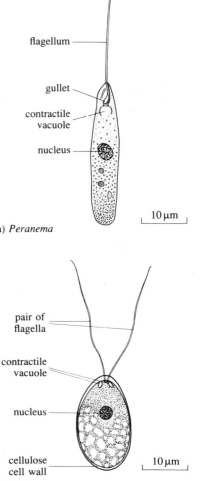

flagellum

gullet

contractile
vacuole

nucleus

10 μm

(a) *Peranema*

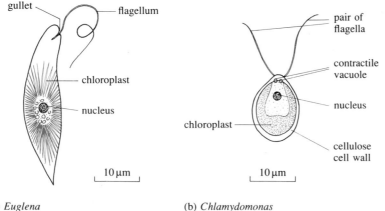

gullet — flagellum

chloroplast

nucleus

10 μm

(a) *Euglena*

pair of
flagella

contractile
vacuole

nucleus

chloroplast —

cellulose
cell wall

10 μm

(b) *Chlamydomonas*

pair of
flagella

contractile
vacuole

nucleus

cellulose
cell wall

10 μm

(b) *Polytoma*

Figure 1.7 Two species of freshwater unicellular algae.

Figure 1.8 Two species of protistans.

algae and cyanobacteria releases a large part of the oxygen in the atmosphere and takes up much of the carbon dioxide generated by the respiration of other organisms and by the burning of fossil fuels.

1.3.3 Multicellular algae

Multicellular algae are primarily aquatic although some intertidal species can survive quite long periods of exposure to air. They normally live attached to rocks and other organisms on and near the shore but detached clumps are often seen floating near the surface. Their common name is seaweeds, although not all multicellular algae are marine. Seaweeds belong to three divisions, the Rhodophyta (red algae), the Phaeophyta (brown algae; Plate 4) and the Chlorophyta (green algae). The last category also includes some unicellular algae (see Section 1.3.2). The basic colour may be masked by that of other pigments so not all members of each group appear to the human eye to be the colour suggested by their name.

Figure 1.9a shows the typical form and internal structure of *Ulva lactuca*, the sea lettuce, a green alga (division Chlorophyta) that is common on British coasts. The alga consists of a sticky holdfast and photosynthetic tissue, which is called the **thallus**, which forms a soft, thin, blade-like structure of variable shape, composed of two layers of chlorophyll-containing cells. Dissolved gases and mineral nutrients are absorbed directly into the thallus cells through their thin, permeable walls. Sea lettuce normally occurs in rock pools and other sheltered places on the sea-shore, where it is firmly attached to the substrate by the holdfast, formed by a group of cells that sticks to hard surfaces. This intertidal alga can withstand several hours of exposure to the air, as well as battering from tides and wave action.

The development and life cycles of algae are varied and in some cases complicated. Figure 1.9b shows the life cycle of *Ulva*. The **sporophyte** plant is diploid and has a holdfast and a thallus, which may form numerous branches. Some of the cells of the thallus divide meiotically, forming single cells that develop two pairs of flagella and swim freely in the plankton, looking and behaving much like unicellular algae such as *Chlamydomonas*, before settling on a suitable rock. Once settled, the spores divide mitotically and grow into a **gametophyte**. In the case of *Ulva* (but not in all algae), the gametophyte is similar in form and size to the sporophyte and also has a thallus and a holdfast.

◇ How does this daughter plant differ from the parent plant that produced the spore?

◆ It is the product of meiosis and is therefore haploid. All its cells contain only half as many chromosomes as were present in the cells of the sporophyte.

Several batches of spores may be released, and a single sporophyte plant can produce thousands of spores and thus many gametophytes. In due course, some cells of the gametophyte thallus divide mitotically and produce motile cells that, in the case of *Ulva*, look and behave much like the free-living spore stage from which their 'parent' plant grew (except that they have only one pair of flagella). However, these gametes swim about in the plankton until they meet another gamete from the same species, with which they fuse to form a zygote before settling and dividing by mitosis to form a sporophyte. Next time the season and environmental conditions are appropriate, some cells of this diploid thallus will undergo meiosis, and the life cycle begins again.

Some of the largest and most elaborate algae are the brown seaweeds (Phaeophyta), including bladder wrack (Plate 4). In many species, the sporophyte is larger and more complex in structure than the gametophyte. A familiar example is kelp, *Laminaria*, which normally forms long fronds attached at one end to a rock or other fixed object. The thallus is divided into a thick, tough stipe, and a thinner, flat lamina which consists of several layers of cells, including some that form filaments running along the length of the lamina, giving it rigidity and toughness. All the cells have cellulose walls and a

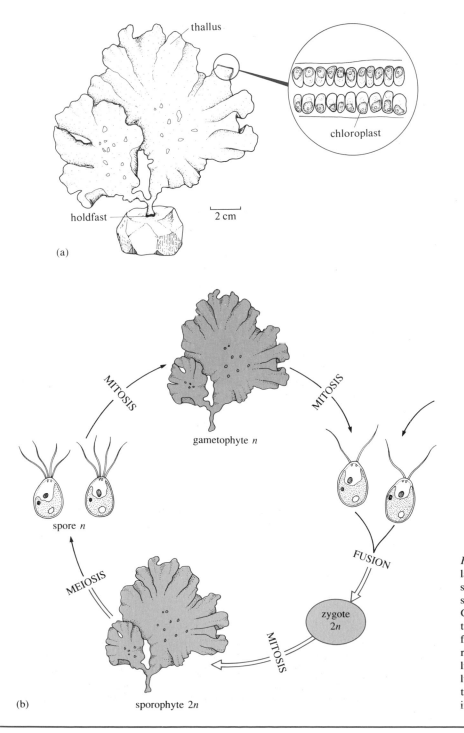

Figure 1.9 (a) A chlorophyte multicellular alga, *Ulva lactuca* (sea lettuce). A section through the thallus shows two sheets of cells. (b) The life cycle of *Ulva*. Gametes are formed from the gametophyte by mitosis; zygotes form from fusion of gametes from the same or different plants. All land plants have similar life cycles although the forms of the thallus and gametes are different. Diploid tissues are shown in pink, haploid tissues in grey. *n* indicates haploid, 2*n* diploid.

thick outer layer of mucilage that protect them from desiccation and from freezing. Although *Laminaria* cannot stand prolonged exposure to air and is therefore confined to the lower shore, some species of brown seaweeds can survive for many days of exposure to air.

Summary of Section 1.3

Eukaryotic plants are classified into seven divisions, and animals into about 25 phyla. There are several distinct groups of animal-like unicellular protistans and plant-like unicellular and multicellular algae. Most protistans are motile and digest food particles in intra-cellular vacuoles. Some have a complex internal structure, including hard skeletons. Protistans are abundant in the soil and in marine and freshwater plankton, and some live in or on multicellular organisms.

All algae are photosynthetic, although some contain red, yellow or brown pigments as well as chlorophyll. The majority have rigid cell walls. Most unicellular algae are planktonic, but many multicellular forms live on or near the shore, attached to fixed objects.

1.4   THE CLASSIFICATION OF TERRESTRIAL PLANTS

All terrestrial plants are in the division Chlorophyta, but they differ from algae (and from fungi) in that there is an embryonic stage in the life cycle, and are hence called the Embryophytina. An **embryo** is a growing, developing multicellular organism that lives surrounded by the tissues of the parent plant and derives nourishment directly from it. The classification of the Embryophytina into classes is based mainly upon their life cycles rather than anatomical features.

There are several basic kinds of eukaryotic life cycles. In protistans, a single mature diploid cell divides by meiosis to form gametes. Two gametes from different parent organisms meet and fuse to form a diploid **zygote** which grows into a new mature organism. The life cycle of most animals is similar, except that the diploid zygote divides many times, forming a multicellular adult organism that consists of many different kinds of cells and may be millions of times larger than the haploid gametes from which it arose. Animal gametes are almost always of two types, the small, motile male sperm and the larger, non-motile female ovum. In terrestrial plants and some multicellular algae (e.g. Figure 1.9b), there is both a multicellular gametophyte (containing haploid cells) and a multicellular sporophyte (containing diploid cells), both of which may consist of several different kinds of cells and have an elaborate internal structure. Instead of arising by meiosis directly from a diploid adult organism (as in animals), the gametes are produced by mitosis in the gametophyte organism, which is already haploid. Such a life cycle is called **alternation of generations**. In the sea lettuce, the haploid and diploid motile stages and thallus are remarkably similar in both structure and physiology, but in this respect *Ulva* is unusual. In most terrestrial plants, the 'generations' live in different environments and hence may be strikingly different in structure and appearance. Many sporophytes that live on land produce spores that are protected from desiccation by a waterproof coat.

Table 1.2 The classification of the division Chlorophyta.

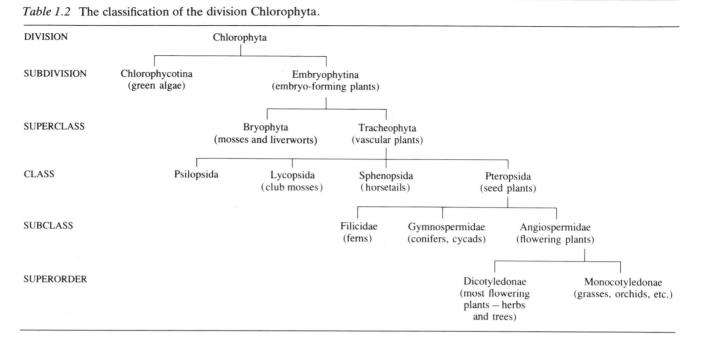

◇ How will the habits of a spore adapted to dispersal on land differ from those of aquatic spores?

◆ A spore that is surrounded by a waterproof coat will be unable to feed and unable to move under its own propulsion.

The spores of most terrestrial plants contain the nutrients required for the establishment of the gametophyte before breaking free from the sporophyte parent plant, and they are often formed and released where they are effectively dispersed by air currents or (in the case of many angiosperms) by animals, particularly insects (see Chapter 2, Section 2.4.3).

◇ Why would similar adaptations be less appropriate for gametes?

◆ Because gametes have to meet and fuse with each other, forming a zygote, before the sporophyte can develop, so they cannot be encased in an impermeable coat.

This section is a brief account of the life cycles and some major anatomical features of the living groups of terrestrial green plants. The scheme for classifying the major groups given in the Appendix and Table 1.2 is only one of several arrangements that have been proposed.

1.4.1 Bryophytes

Bryophytes are liverworts (Plate 5a) and mosses (Plate 5b), the simplest of the terrestrial green plants. They lack efficient internal structures for support and transport of water and nutrients within the plant. Mosses are low-growing plants, usually with leafy shoots that grow in a regular pattern; liverworts typically have a flat, sheet-like growth form. They do not have roots or flowers like those of the more familiar seed plants (see Section 1.4.3). Although some mosses that are adapted to seasonal drought can survive long periods in a dry dormant state, the gametes of bryophytes are aquatic and they do not form seeds, so they need standing water to complete their life cycle. Bryophytes are generally inconspicuous plants that do not grow under prolonged exposure to extremes of dryness, heat or cold. Most are confined to damp, usually shady habitats where there is a film of water for at least part of the year. Nonetheless, they are, and have been since plants first appeared on land, moderately abundant in tropical, temperate and arctic climates.

Figure 1.10 shows the general scheme for the life cycle of the moss *Mnium*, with illustrations of the structures of the principal stages. The most familiar stage, that of the plant commonly called 'a moss', is the gametophyte. It has two different gamete-producing structures that form larger, non-motile female gametes and tiny flagellated male gametes. Normally, the male gametes are produced in very large numbers and swim in water, guided by chemical attractants secreted by the female reproductive organs. A layer of water is essential for successful meeting and fusion of the gametes. Most gametophyte mosses can only reproduce in places where a film of water forms on a surface or near semi-permanent water such as ditches and bogs, and the gametes of those living in drier places are often released during showers of rain.

After fertilization, the zygote develops into a diploid sporophyte. Moss sporophytes remain attached to the gametophyte plants on which they are dependent as a source of food. The sporophyte has tougher supporting tissues than the gametophyte, and most of its surface is impermeable to water and to gases, which pass through apertures called **stomata** that form on the shoots. Capsules develop on the top of the stalks and bear superficial resemblances to tiny fruits. Meiosis takes place inside the capsule and the haploid spores are released, usually following drying out of the capsule.

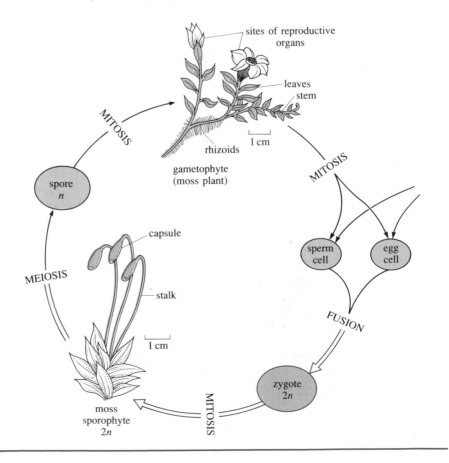

Figure 1.10 Life cycle of a bryophyte moss, *Mnium*. Diploid tissues are shown in pink, haploid tissues in grey. Gametes are formed from the gametophyte by mitosis; zygotes form from fusion of gametes from the same or different plants. *n* indicates haploid, 2*n* diploid.

1.4.2 Ferns and fern-like plants

Several different kinds of plants are grouped together as the Tracheophyta ('vessel plants'; see Table 1.2) because they all have special cells that form 'vessels' which convey water and mineral nutrients through the plant. All these plants (except the rare psilopsids) have stout, leafy shoots and true roots which in many cases grow from a horizontal underground stem called a rhizome. Ferns (Plate 6c) and fern-like plants (Plates 6a and 6b) never have flowers or seeds and, like bryophytes, their life cycle includes an aquatic stage and so cannot be completed without water. They are widespread and in some places abundant on land, and many grow to the size of a small tree.

The life cycle of typical ferns such as *Dryopteris* (Figure 1.11) is basically similar to that of mosses. A motile male gamete and a protected sessile egg are produced by the gametophyte generation, and fertilization takes place in water. However, the most conspicuous and familiar stage of the fern life cycle is the sporophyte, which is well able to feed itself from its tough, erect green leaves. The sporophytes of some ferns, for example *Pteridium aquilinum* (bracken), thrive in dry conditions where the diploid stage spreads by vegetative reproduction. However, the gametophytes of ferns lack the impermeable outer covering and tough **vascular system** and are much less able to live away from water. The gametophyte of *Dryopteris* is a small thallus with simple root-like rhizoids, but nothing resembling leaves. In more advanced ferns, the gametophytes may be even further reduced to simple multicellular spores that are dispersed by wind. The female gametes remain attached to the gametophyte, while the male gametes are tiny, encapsulated

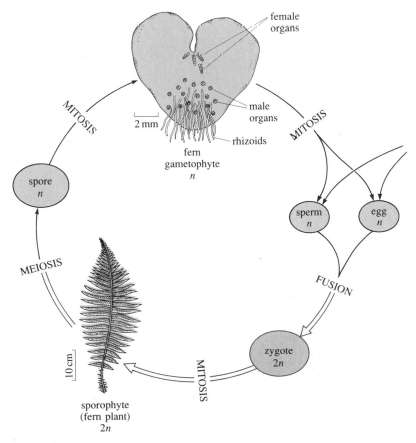

Figure 1.11 Life cycle of a fern, *Dryopteris felix-mas*. Diploid tissues are shown in pink, haploid tissues in grey. Note that the gametophyte is seen from below. Gametes are formed from the gametophyte by mitosis; zygotes form from fusion of gametes from the same or different plants. *n* indicates haploid, 2*n* diploid.

particles which release motile sperm cells and fertilization takes place only in water. In some advanced forms, the female gametophyte and the female gametes may remain attached to the sporophyte.

The three most familiar kinds of fern-like plants are the Lycopsida (Figure 1.12a and Plate 6a), commonly called 'club-mosses', although they are fundamentally different from the true mosses described in Section 1.4.1, the Sphenopsida (horsetails; Figure 1.12b and Plate 6b) and the Filicidae, the true ferns, including *Polypodium* (Figure 1.12c). Most fern-like plants are confined to damp, usually shady habitats where films of water form for at least part of the year.

1.4.3 Seed plants

All large terrestrial plants are tracheophytes that bear seeds. There are two principal categories of seed-bearing plants (Table 1.2): the Gymnospermidae ('naked seeds'), which include the conifers (Plate 7a) and less familiar trees such as ginkgos and cycads, and the flowering plants Angiospermidae ('closed seeds'). The flowering plants are divided into the monocotyledons (Plates 7e and 15f), e.g. grasses, bamboo, rushes, irises and orchids, which have narrow leaves with parallel vessels, and the much more diverse broad-leafed dicotyle-

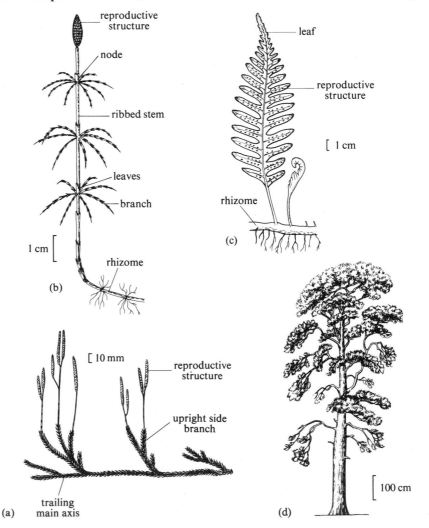

Figure 1.12 Land plants (a) A club moss, *Lycopodium clavatum* (Lycopsida). (b) A horsetail, *Equisetum sylvaticum* (Sphenopsida). (c) Fern, *Polypodium vulgare* (Filicidae). (d) Pine tree, *Pinus sylvestris* (Pteropsida, Gymnospermidae).

dons (most shrubs, herbs and trees, Plates 7b–d), in which the leaf vessels branch, often forming an elaborate pattern. The two basic components of the sporophyte of seed plants are the roots, through which the plant draws water and nutrients from the substratum, and shoots, which include the stems, leaves and reproductive structures, that are exposed to light and air. Chloroplasts may occur in all kinds of shoot cells, including stems and reproductive tissues but in most larger seed plants, particularly trees, photosynthesis occurs mainly in the leaves, which contain most of the chloroplasts and are specialized for absorbing light and taking in carbon dioxide. Seed-bearing plants also have efficient vascular systems that transport nutrients and water through the roots, stems and leaves. Many seed-bearing plants, including all large forms, contain wood, a structural tissue that consists mainly of dead (i.e. non-metabolizing) cells and some of which form tough, flexible fibres interwoven to form a rigid load-bearing material (Figure 1.12d).

As their name implies, one of the most important characteristics of these plants is the formation of enclosed seeds that, in many species, are remarkably resistant to damage from weather and other organisms, remaining viable for much longer than the embryos of animals or the spores of bryophytes or fern-like plants. Seed plants (except those secondarily adapted to living in water, e.g. water-lilies) can complete their entire life cycle in the absence of standing water. There is no free-living gametophyte generation: the zygote forms on the sporophyte plant and becomes part of an enclosed seed.

In angiosperms, the formation of the gametes and fertilization take place in the **flower**, which is a modified shoot of the diploid sporophyte that forms the **ovary** and structures that disperse and collect the **pollen**. Flowers (Plates 7b–e) include structures that nurture the maturation of the ovule (female gametophyte containing the female gametes) and the fertilized zygote and structures that release the pollen. Pollen is a spore stage that remains dormant until contact with the female structures triggers germination to produce the male gametophyte, one nucleus of which becomes the male gamete. The basic structure of a typical angiosperm flower is shown in Figure 1.13. When mature, the flower contains pollen, which is enclosed in a tough

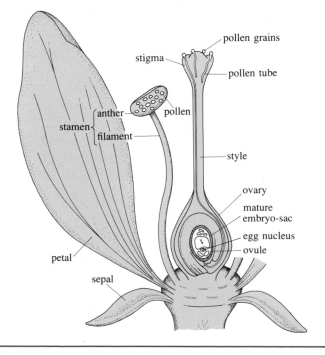

Figure 1.13 The structure of a typical angiosperm flower. Diploid tissues are shown in pink, haploid tissues in white.

outer covering, and may be dispersed by wind, or, as in many angiosperms by pollinating animals (see Chapter 2, Section 2.4.3), and the embryo-sac and ovule remain attached to the parent sporophyte. The pollen grains that transfer the male gamete to the female reproductive structures are non-motile and, being protected by an impermeable coat, are very durable, some remaining viable for tens of years. The male and female reproductive structures may be on the same flower, or on different kinds of flowers on the same plant (Plate 7b) or, in a minority of species, on different plants.

Events leading to fertilization begin when pollen, usually originating from another plant and transmitted by wind (Plate 7b) or pollinating animals (Plates 7c–e), falls on the stigma (Figure 1.13). The pollen grain germinates to form the male gametophyte which elongates and forms the pollen tube which grows down the style towards the female gametophyte. The tiny, fragile male gametes, which are simply nuclei, then move down the enclosed, protective pollen tube and fertilization takes place in the ovary. One male gametic nucleus enters the ovule and forms a zygote which develops into a diploid embryo. A second pollen nucleus fuses with two others and the resulting triploid cell divides to form a storage tissue called the endosperm.

The ovary then produces a tough, impermeable, sometimes woody coat around the embryo and endosperm. In most angiosperms, mitotic cell division of the zygote begins shortly after fertilization but development of the embryo is arrested while it is still very small. The tissues combine in the ovary to form the **seed**, which consists of the tiny embryo, the **endosperm**, which will provide nutrients when embryonic development restarts after germination, and one or more layers of protective covering (Figure 1.14). The ovary may also produce one or more layers of nutritive or protective tissue around each seed or group of seeds, forming **fruits** or pods. The fruits also promote dispersal of the seeds either by wind (e.g. sycamore, dandelions, most grasses) or by animals (e.g. tomatoes, acorns). The whole process of formation of the seeds and fruit may take from many months (e.g. apples, avocados) to a few hours (e.g. dandelions). Protected by their stout coat, seeds may remain dormant but viable for months or years until they germinate and the development of the embryo is re-activated. Germination is often triggered by increases in temperature and moisture, changes in the environment that indicate that the weather is best suited to the growth of a fragile young plant. Once germinated, seeds cannot return to dormancy.

The gymnosperm life cycle is similar to that of angiosperms except in the following ways:

1 The gametophytes and seeds form on cones instead of flowers.

2 Both male and female gametophytes are relatively larger and more complex in gymnosperms.

3 The pollen is dispersed by wind and the pollen tube liberates motile male gametes.

4 The seed is 'naked' because the ovary does not form a protective coat around it.

5 Some of the more primitive gymnosperms are pollinated by animals, but all conifers are wind-pollinated and the reproductive structures that develop on the sporophyte are rarely brightly coloured or attractively scented (see Chapter 2, Section 2.4.3).

The presence of flowers and seeds eliminates from the life cycle of gymnosperms and angiosperms an aquatic stage such as the swimming gametes of bryophytes and ferns. Seed-bearing plants are by far the most abundant and diverse elements of the contemporary terrestrial flora. In fact, they are so effectively adapted to live on land that there are very few completely aquatic

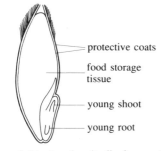

protective coats

food storage tissue

young shoot

young root

Figure 1.14 A longitudinal section through an oat seed. The bristles on the protective coat aid dispersal of the seed.

species. Some exceptions are *Zostera* (eel-grass) in the tidal zone of temperate seas (including those around Britain), plants that live in shallow freshwater ponds and lakes, such as water-lilies, and mangroves in tropical estuaries and tidal mud flats.

1.5 FUNGI

Yeasts (Figure A.10), mushrooms and toadstools (Figures A.12 and A.14 and Plate 8b), moulds (Figure A.9), mildews and the 'rusts' and 'smuts' (Figure A.13) that infect higher plants are fungi (Figure 1.15). Fungi have some characteristics in common with protistans (e.g. slime moulds, Figure A.8) but also have a number of unique features that have not been described in any other organisms, living or fossil. Terms and concepts derived from observations of plants and animals are often invalid for fungi and you should bear this in mind when thinking or writing about them.

Most fungi (Figure 1.15) contrast with filamentous protistans and algae and with other multicellular organisms in that the cells form **syncytia** ('together cells'), consisting of scores of nuclei and large quantities of cytoplasm enclosed in the same cell membrane, and surrounded by a cell wall. Like multicellular chlorophyte plants, fungi are non-motile organisms consisting of living tissue enclosed in tough cell walls but, unlike green plants, they never contain chlorophyll and do not photosynthesize. Fungi are heterotrophic but, instead of 'eating' other organisms whole or in large pieces as most animals and protistans do, they absorb nutrients from their surroundings. The majority are **saprophytes** ('putrid plant'), feeding on the tissues of dead or dying plants and animals on land or in freshwater. Only a very few kinds of fungi can survive in seawater.

Phycomycetes are structurally simple fungi. They have an aquatic phase in the life cycle and the hyphae and cell walls are simple in form and they normally cannot break down cellulose or other complex carbohydrates. Some familiar examples are the bread mould and potato blight. The yeasts and some moulds (including *Penicillium*) and some toadstools such as morels are members of the class Ascomycetes. The smaller species consist of single cells or strings of identical small cells, loosely connected to form an amorphous mass. They can break down a wide range of complex carbohydrates. The most elaborate of fungi are the class Basidiomycetes which have complicated life cycles and can

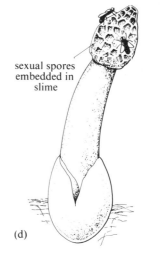

sexual spores embedded in slime

(d)

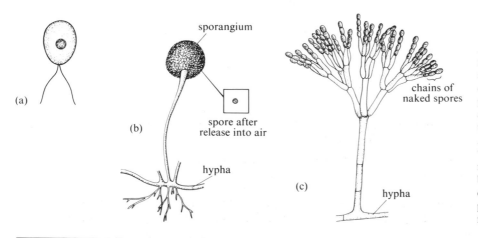

(a)

sporangium

(b)

spore after release into air

hypha

(c)

chains of naked spores

hypha

Figure 1.15 Reproductive stages of various fungi. (a) The flagellated spore of an aquatic phycomycete. (b) Wind-dispersed asexual spores of a terrestrial phycomycete. The spores are enclosed in a fruiting body (the sporangium) until they are released. (c) An ascomycete, *Penicillium*. The asexual spores are 'naked' (i.e. not in capsules) and are released singly from chains. (d) A fruiting body of a structurally elaborate basidiomycete, the stinkhorn. This fungus produces a stink that attracts carrion-feeding flies, shown here to provide scale.

often break down wood. Some of the most familiar fungi are large basidi-omycetes, including the edible mushrooms (Plate 8b) and most toadstools, but there are also some smaller species of economic importance such as rusts and smuts, which damage wheat and other crops.

These fungi consist of thread-like **hyphae** (singular: hypha, 'web'), usually white in colour, that grow only at the tips and are often variable in form. When examined with the light microscope (Figure 1.15c), hyphae appear as transparent, tube-like structures with rigid walls packed with living material containing numerous tiny nuclei. The living material is thus not always divided into discrete cells, although many hyphae are divided into interconnecting chambers with partitions called septa and the gametes are single cells with one nucleus. Fungi cannot take in large particles by phagocytosis because the cells are surrounded by a tough cell wall. Instead, the hyphae secrete enzymes that break down the food so 'digestion' takes place externally. The fungus then absorbs the complex organic molecules derived from dead organisms and inorganic nutrients through the cell walls that are permeable only to relatively small molecules. As with bacteria and protistans, simplicity of anatomical form does not necessarily mean simplicity of biochemical capacity; fungi have many complex and unique biosynthetic mechanisms and can obtain nutrients from materials such as wood much more efficiently than most animals.

Most fungi form spores, which can remain dormant for long periods and are transported by flowing water, air currents or on motile animals. The spores of simple fungi are usually diploid, but in the more complex species, haploid spores form on sporangia (singular: sporangium), which protrude out of the soil or decaying organism on which the fungus is feeding. Indeed, the most familiar and spectacular fungi are the fruiting bodies of large forms such as mushrooms and toadstools, which have an elaborate internal structure and complex external form (Plate 8b). It is important to remember that their dormancy and capacity for dispersal are the only points of similarity with the 'spores' of bryophyte and tracheophyte plants (Section 1.4): fungal spores are diploid or haploid, while most plant spores are haploid.

When a spore settles on a suitable source of food, it sheds its protective coat and begins to develop hyphae (Figure 1.15) that spread over and through the food source, forming a mass called the **mycelium** (plural: mycelia; Plate 8a). The thin-walled hyphae of most fungi are susceptible to desiccation, and so grow best in warm, damp conditions. However, in moist, sheltered places, such as soil and in the tissues of dead organisms, and on abundant, nutritious food, fungi can grow impressively fast. You will be familiar with the many kinds of moulds and mildews that grow on ripe fruit, bread, cheese, meat and many other foods, particularly if they are kept in warm, damp places. Only a few fungi that have thickened, fibrous hyphae covered with a waterproof coating can grow in dry environments; one such species is the dry-rot fungus *Serpula lacrymans*, that feeds on wood, destroying old timber in forests and, unfortunately, in buildings.

Summary of Sections 1.4 and 1.5

Terrestrial plants are classified mainly on the basis of life cycle and the structure of vessels that transport fluids within the plant. Alternation of generations means that the plant or alga has a multicellular haploid phase as well as a multicellular diploid phase. In animals (and some algae) the only multicellular stage is normally diploid. In bryophytes, ferns and fern-like plants, the gametes are aquatic so the life cycle cannot be completed without contact with free water. The most elaborate stage of moss plants is the gametophyte, which has no roots or internal vessels, but the most massive stage of ferns and fern-like plants is the sporophyte which has internal vessels. Seed plants have internal vessels and the fertilization and development of the gametophyte take place on the sporophyte generation and do not require free water.

Fungi are heterotrophic, non-motile organisms many of which form syncytia and have many structural and biochemical features not found in any other eukaryotes. Their basic structure consists of syncytia that form hyphae. They feed mainly on dead plant and animal remains and are abundant in soil.

Question 4 (*Objective 1.5*) Figure 1.16 shows four kinds of life cycle.

(a) Which type of life cycle occurs in both algae and land plants?

(b) Which type of life cycle resembles that found in animals?

(c) Which life cycles show alternation of generations?

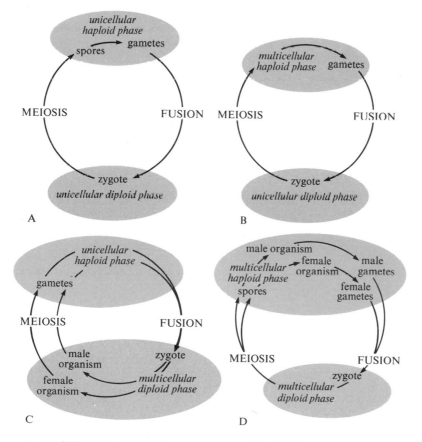

Figure 1.16 Types of life cycles (for use with Question 4). Diploid tissues are shown in pink, haploid tissues in grey.

Question 5 (*Objective 1.6*) Suggest a group in which each of the organisms described in (a)–(c) should be classified.

(a) A multicellular, flattened green thallus with cells that have distinct cell walls, nuclei and mitochondria.

(b) A flattened colony consisting of chains of cells with rigid cell walls but without growing tips, or visible chloroplasts, nuclei or mitochondria.

(c) A flattened colony consisting of chains of cells with rigid cell walls, all growth being at the tips of the colony; nuclei and mitochondria are visible at the tips, but chloroplasts are absent.

1.6 THE CLASSIFICATION OF ANIMALS: BODY PLANS AND LIFE CYCLES

One of the most important criteria for identifying and defining animal phyla is the anatomical arrangement of the major tissues and organs: the basic body plan. Like protistans (Section 1.3.1), animals are heterotrophic, taking in relatively large particles of food. In typical animals, the food is then digested and absorbed through the wall of the gut and transported from there to the other tissues of the body, often in a fluid tissue, called **blood**, that circulates between the organs and perfuses them with respiratory gases and nutrients. Blood usually contains large numbers of living cells as well as dissolved gases and nutrients and substances **secreted** by other tissues (e.g. hormones). Nutrients are metabolized to release energy usually in conjunction with oxygen absorbed through the body surface, or through special **respiratory organs**. The other specialized tissues include excretory organs, which collect and eliminate waste, and muscles, which enable the animal to move and thereby to obtain its food. The muscles are the major consumers of the nutrients and respiratory gases, and the arrangement of the gut in relation to the skeleton, muscles and other tissues is an important factor in determining the habitats and kinds of food sources that the animal can exploit. Sense organs detect chemicals in the environment (smell and taste), vibration (ears and mechanoreceptors), light (eyes) and, in some fishes, electric currents. The activities of all these organs are coordinated by the nervous system which has connections with the muscles, gut and respiratory organs. In some animals (e.g. vertebrates) the skeleton is also a specialized tissue, but in many other groups, the main structural material is acellular. An animal's body plan is a long-standing and important criterion for distinguishing major phyla of animals and can be studied quite easily by dissection of whole specimens or by examination of sections.

Another criterion for the classification of animals is their life cycle. In many animals, the body form, diet and habits of the immature stages (**larvae**; singular larva) are different from those of the adults. Sometimes, these larvae are so unlike their parents that the only way of finding out what they will become when they grow up is to rear them in captivity and see! Nonetheless, larval body plans are often as important as those of adults as indicators of ancestry and taxonomic affinity (see Chapter 2, Section 2.2.1).

1.6.1 Porifera

Poriferans ('pore-bearing') are an ancient but still fairly abundant group of animals, commonly known as sponges. The great majority live in the sea, often attached to rocks, driftwood or the shells of other animals, although a few species live in freshwater. Their basic structure is illustrated in Figure 1.17 and Plate 9. Sponges extract their food from a current of water that enters through a variable number of inhalant pores, scattered over the surface; there is normally only one or a few oscula (sing. osculum), through which the water is expelled. There are three main kinds of cells, the flagellated choanocytes, which normally beat synchronously, generating the water current and taking in minute particles of food, a second type that secretes the spicules of the skeleton, and a third type that forms the outer layer. Sponge cells can change form: there are numerous cells called amoebocytes because they resemble *Amoeba* (Figure 1.5d) that wander through this matrix, or may become gametes and break away from the parent sponge. At any one time, however, the numbers of each type of cell are more or less constant. Sponges grow irregularly, depending upon the availability of

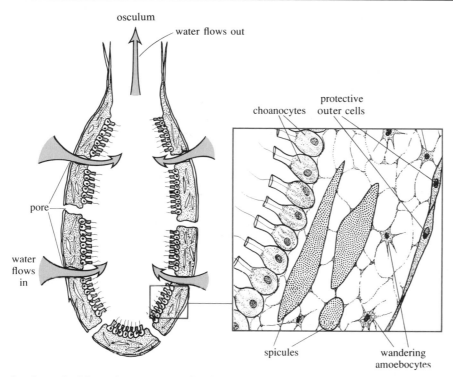

Figure 1.17 Diagram of a simple sponge with part of the wall enlarged to show the types of cells. The arrows indicate the direction of water currents.

food, and although many species have a more or less consistent shape, and some become quite large, they do not have any clear symmetry. The natural bathroom 'sponge' is the dried remains of the extracellular protein (called spongin) that in the living animal connected the spicules together.

1.6.2 Cnidarians (coelenterates) and ctenophores

Figures 1.18a, b and c, 1.19, 1.20 and 1.21 are sections of five different animals. Tissue in which cells are few or absent are shown as black or grey tone.

◇ Which animals consist of only two layers of cells? In such forms, where is the gut in relation to the cellular tissues?

◆ The animals shown in Figures 1.18a, b and c. There are two layers of cells separated by a non-cellular **mesogloea**. The gut (called the enteron) is in the centre of the body, lined by endoderm.

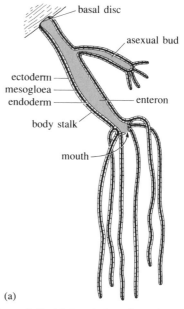

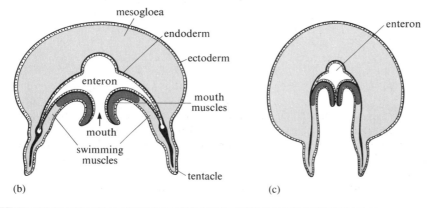

Figure 1.18 (a) Vertical section of simple coelenterate polyp, *Hydra*. (b) Vertical section of jellyfish with the circular swimming muscles relaxed. (c) Vertical section of jellyfish with the circular swimming muscles contracted.

Figure 1.18a is a section of a *Hydra* polyp and Figures 1.18b and c are vertical sections through a jellyfish. Both animals are classified as Cnidaria ('stinging cells'), a large and diverse phylum characterized by having only two layers of cells, an inner **endoderm** ('inner skin'), an outer **ectoderm** ('outer skin') and a unique and elaborate kind of cell, the **cnidoblast** ('nettle cell'), that contains long threads, often armed with barbs or toxic chemicals. The cnidoblasts are concentrated on the tentacles and are discharged by chemical or mechanical stimuli. The toxins immobilize small animals such as crustaceans and fish larvae; those of some large jellyfish are lethal to much bigger animals, including humans. All cnidarians are basically **radially symmetrical** and the gut has only one opening (Figure 1.18). Both the endoderm (inner layer) and the ectoderm (outer layer) consist of several different kinds of cells. Some cells of both the endoderm and the ectoderm contain muscle fibres, many of which are capable of fast and powerful contraction. Their activity, and that of the cnidoblasts, is coordinated by cells that can propagate electrical signals. Many such cells are elongated and when examined under the microscope, look similar to neurons (see Chapter 3, Section 3.2.1) in the nervous systems of other phyla, but other conducting cells are cuboidal in shape and may have other functions besides conduction.

Cnidaria (also called coelenterata, 'hollow guts') occur in two principal body shapes, the tubular **polyps** (e.g. sea-anemones, corals and *Hydra*, Figure 1.18a) which are normally sessile, and the bell-shaped **medusae** (Figures 1.18b and c and A.18), many of which live in the open ocean and swim actively (e.g. jellyfish).

Most cnidarian polyps can wave and shorten their tentacles and withdraw them as the mouth is closed. Anemones (Figure A.19) are single polyps with a complex internal structure formed by folds of endoderm. The polyp *Hydra* is just a small, simple sac that lives in freshwater pools, attached to water plants, and feeds on small animals. For these reasons it is convenient for biological research, and has been cultured in laboratories for many years. However, *Hydra* and a few other cnidarians are unusual in living in freshwater: the great majority of others are marine. Cnidarians occur in all areas of the sea, from sea-anemones which are in the intertidal zone of the shore and exposed regularly to corals (Figure A.20) in the deep sea. Most medusae swim by a characteristic pulsating action. The powerful, synchronous contraction of the circular muscles bends the bell and deforms the elastic, jelly-like mesogloea, ejecting a jet of water from under the bell, which propels the animal. When the muscles relax, the mesogloea springs back to its former shape. The mesogloea consists mainly of fibrous proteins and water and contains only a few scattered cells, none of which is organized into organs.

◇ How do the requirements of cells differ from those of acellular tissues such as mesogloea?

◆ Living cells, but not acellular materials, need access to oxygen and nutrients, and a means of eliminating waste.

◇ How could these different requirements affect the anatomical arrangement of cellular and acellular tissues?

◆ Living cells must be in contact with the external medium or with body fluids such as blood. Acellular tissues can exist in large blocks without contact with such nutrient-supplying media.

Consequently, in spite of their simple body plan, many cnidarians become large. Huge jellyfish, several metres in diameter, are some of the largest invertebrate animals in the plankton. Some polypoid and medusoid cnidarians form colonies that may be elaborate and extensive. Colonial coelenterates include the reef-building corals (see Chapter 2, Section 2.4.1) and floating forms such as *Physalia*, the Portuguese man o' war. Some kinds of hard corals form colonies that can extend over large areas, and the trailing tentacles of *Physalia* can be several metres in length.

The Ctenophora ('comb bearing'), commonly known as sea-gooseberries or comb jellies, are similar to cnidarians in many ways. Like cnidarians, they have two layers of cells and a largely acellular mesogloea, but they lack cnidoblasts. They are all marine and most live in the plankton, preying on smaller animals, and swim with the rows of cilia from which they derive their name. Although the group includes relatively few species, some of them are very abundant in temperate and arctic oceans.

1.6.3 'Worms'

The term 'worm' means any long, thin, soft-bodied animal that normally moves by wriggling or crawling. Flatworms, roundworms, bootlace worms, earthworms, many caterpillars, certain beetle and fly larvae, sea-cucumbers, eels, snakes and other legless vertebrates are all worm-like, but detailed examination reveals fundamental contrasts in their internal structure.

◇ Compare Figures 1.19 (a planarian) and 1.20 (a roundworm) with Figures 1.18a and b. Name some similarities and contrasts in the body plans.

◆ In both cnidarians and these worms, continuous layers of cells form the outer covering (ectoderm) and the lining of the gut (endoderm). In the cnidarian and the flatworm, but not the roundworm, the gut is a sac with a single opening.

In place of the largely acellular mesogloea, there is a cellular **mesoderm**. The majority of tissues, including muscles, internal skeleton and excretory organs, are formed from the mesoderm. These worms also differ from cnidarians in several other important ways: they are bilaterally rather than radially symmetrical, and they never have cnidoblasts.

Platyhelminths

The phylum Platyhelminthes ('flat worms') (Figures 1.19 and A.21 and A.22) are common in freshwater and in the sea, and a few live in damp soil and leaf litter. Most of the free-living forms belong to the class Turbellaria (the name means 'little disturbance', a reference to the fact that sometimes dozens of

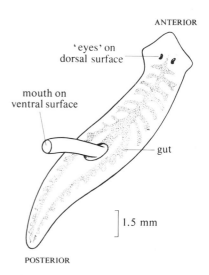

(b)

Figure 1.19 Typical free-living platyhelminth. (a) A cross-section through a flatworm, with the principal muscles of the mesoderm shown in red. (b) Diagram of a whole flatworm showing the gut with the pharynx everted through the ventral mouth and the 'eyes' at the anterior end.

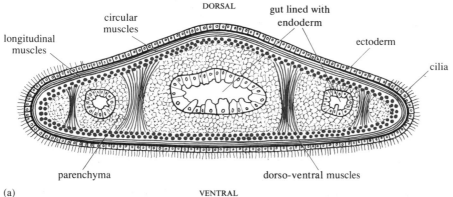

(a)

them gather at a source of food). They are semi-transparent worms with light sensitive spots or 'eyes' at the 'head' end and cilia over most of the ventral surface. The mouth is on the end of an extensible **pharynx** (Figure 1.19b) that protrudes from the middle of the ventral surface of the body. The gut forms several branches and numerous blind pockets (Figure A.21), but there is no anus, undigested material being evacuated through the mouth. The mesoderm contains numerous bands of muscles with which the worm can twist, flatten and lengthen its body, enabling it to squeeze through crevices and also to 'crawl' by passing ripple-like waves of contraction along the ventral surface. Turbellarians also glide across stones or vegetation by beating the cilia on the ectoderm of the underside of the body (Figure 1.19a). The loose, cellular **parenchyma** performs a variety of functions, including elimi- nating waste products and supporting the gonads. The nervous system consists of several small, poorly defined nerve cords and a diffuse network of conducting cells that links the eyes and other superficial sensory cells with the muscles. Most free-living flatworms, called planarians, are carnivorous and suck in small animals and scraps of carrion through the muscular pharynx.

The majority of platyhelminths are parasites; the flukes belong to the class Trematoda ('perforated') worms and resemble the turbellarians in general body form, but the tapeworms (class Cestoda; 'like a belt') are very different in form from their free-living relatives. Flukes and the larval stages of tapeworms are usually small organisms that live in the liver, blood system or muscle of their hosts. Adult tapeworms typically live in their host's gut, usually attached at the anterior end. The body grows by adding sections, each of which contains most of the important organs, and some species may reach many metres in length. Flukes and tapeworms are some of the most physiologically specialized and economically important animal parasites (see Chapter 2, Section 2.4.2).

Nematodes and annelids

The body plans of roundworms (Figures 1.20 and A.24) and of earthworms and ragworms (Figures 1.21 and 1.22 and Plate 10a) are also regarded as sufficiently different from each other and from the platyhelminths for them to be placed in the phyla Nematoda ('like a thread') and Annelida ('ringed')

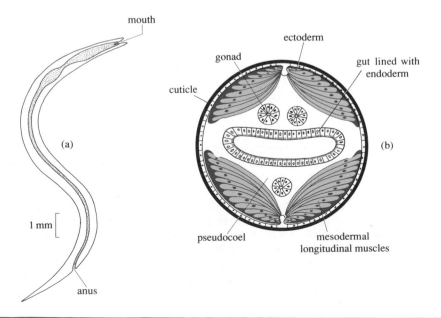

Figure 1.20 A typical free-living nema- tode. (a) Diagram of a whole round- worm. (b) A transverse section through the body of a roundworm. The muscle fibres are shown in pink.

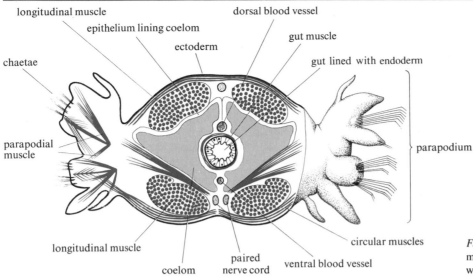

Figure 1.21 Cross-section through the middle of the body of a typical annelid worm.

respectively. (In some older texts, nematodes and some other less familiar groups of worms were classified together as the phylum Aschelminthes.) The body plans of both these phyla have some obvious similarities to that of platyhelminths, including cellular ectoderm, endoderm and mesoderm. The organization of the nematode nervous system is similar to that of platyhelminths. However, the gut of annelids and nematodes has two openings, an anus as well as a mouth, and there are one or more body cavities. The form and cellular lining of this cavity within the mesoderm is one of the main criteria used to distinguish the nematodes from the annelids. In annelids, there is a fluid-filled cavity, lined by epithelial cells and situated between the muscles around the gut and the locomotory muscles. The mesoderm is thus split into an inner and an outer layer. This arrangement of body cavity is called a **coelom** ('cavity'). Cellular tissues such as reproductive organs may develop from either of the two layers of mesoderm and may protrude into the cavity. Body cavities with similar anatomical relations to other tissues are found in several other phyla.

◇ Compare Figures 1.20 and 1.21 and name two ways in which the arrangement of the body cavity and muscles differ in annelids and nematodes.

◆ There is no muscle around the gut of nematodes and the body cavity lies between the endodermal gut wall and the mesodermal locomotory muscles. The nematode body cavity is not lined with epithelial cells.

For these reasons, the body cavity of nematodes is called a **pseudocoel** ('false cavity'). The fluid in the coelom or pseudocoel (or in the case of platyhelminths, the parenchyma tissue) acts as a **hydrostatic skeleton**, supporting the body at rest and during movement. In nematodes, the longitudinal muscles contract alternately, deforming the stiff outer cuticle (Figure 1.20b) and the hydrostatic skeleton, producing a characteristic whip-like movement. In soft-bodied worms, such as annelids (Figure 1.21), contraction of the circular muscles squeezes the coelom and makes the body thinner. The volume of the body is constant since tissues and coelomic fluid do not normally escape, so the body will become longer as it narrows. Conversely, if the longitudinal muscles contract, the body will become shorter but thicker. If part of the body is attached to or pressed against a substratum, lengthening the body pushes

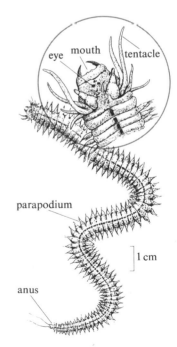

Figure 1.22 A polychaete worm, *Nereis* (ragworm), seen crawling from above. Note the numerous similar segments each with a pair of parapodia. The enlarged head shows the jaws and the eyes.

the front forward. Adhesion of short, broad regions of the body onto the substratum is often aided by bristle-like chaetae that protrude from the body wall. Thus a soft-bodied worm can crawl along or through mud, or swim by alternating contraction of the circular and longitudinal muscles.

◇ Compare Figures 1.21 and 1.22 with Figures 1.19 and 1.20 and name two other contrasts between the basic body plans of platyhelminths and nematodes and that of annelids.

◆ The body of annelids is divided into **segments**; there are **blood vessels** and there is a paired nerve cord. All these features are absent in platyhelminths and nematodes.

The anterior region of the dorsal blood vessel is thicker and more muscular; this and lateral vessels actively pump the blood and so act as a 'heart'. In many annelids, including the familiar earthworm and lugworm, the blood contains pigments that combine reversibly with oxygen. One of the most widespread annelid blood pigments is the iron and protein complex called **haemoglobin**, which is remarkably similar in structure and properties to the haemoglobin of vertebrates. Therefore, the tissues of some annelids, like those of vertebrates, appear bright red.

In typical annelids, the body wall is soft and flexible, the coelom is partitioned by **septa** (Figure A.29) and the longitudinal muscles are divided into blocks along the body. The coelomic fluid of annelids differs from the blood in that it contains few cells, never has respiratory pigments and it is not actively pumped around the body by a specialized organ.

◇ How would dividing the coelom into a series of cavities affect the mechanism of locomotion?

◆ In a non-partitioned coelom, muscular contraction anywhere could alter the shape of the whole body. If the coelom is divided by septa, contraction of the circular and longitudinal muscles would have most effect on the shape of their own segment.

Although the septa are soft and deformable, so some of the pressure changes are transmitted to adjoining segments, their presence permits much more precise movements than are possible for worms with an undivided hydrostatic skeleton.

There are three major classes of annelid worms: the Polychaeta ('many bristles'; Figures A.28a and b), most of which are marine carnivores or filter feeders (Plate 10a), Oligochaeta ('few bristles'; Figure A.29), which feed mainly on plant material or detritus, and Hirudinea (leeches; Figure A.30). Some leeches are predators on other soft-bodied invertebrates, but the majority are bloodsuckers, attacking vertebrates such as fishes and mammals and large invertebrates. Both oligochaetes and leeches occur in freshwater, in the sea and on land, but are most abundant and diverse in ponds, lakes and slow-flowing rivers and in moist terrestrial habitats such as soil and rainforest.

A widespread and familiar oligochaete is the earthworm *Lumbricus*. Nutrients, respiratory gases and other essential supplies are actively pumped in the blood system (Figure A.29) to internal organs in the mesoderm, particularly the large locomotory muscles (Figure 1.21). The nervous system consists of a pair of cords running the length of the body, and clumps of nerve cells called ganglia in each segment that coordinate muscular activity within and between segments. The presence in annelids of a coelom, a blood vascular system, a centralized nervous system, and the segmental arrangement of these organs and the musculature, make locomotion using bands of muscle and a hydrostatic skeleton much more powerful and efficient than it is in flatworms or roundworms.

The movements of earthworms during crawling and burrowing have been thoroughly studied: one part of the body becomes short and thick, and so is pressed against the substratum, while adjoining segments are elongated and lift off the substratum. You can observe waves of shortening and elongation in a common earthworm (Figure A.29) placed on a hard surface or in a narrow glass tube, and when you let it crawl between your fingers, you can feel its chaetae. Contraction of the longitudinal muscles on only one side of a segment produces side-to-side movements. Most polychaetes have segmentally arranged pairs of appendages called parapodia (sing. parapodium) that contain bands of strong muscles and terminate in stout chaetae. Parapodia act as paddles in swimming or burrowing. Many polychaetes such as the ragworm *Nereis* (Figures 1.21 and 1.22) are active carnivores that pursue prey through mud, sand and dense vegetation in shallow seas.

◇ Consider the structure and arrangement of tissues around the guts (Figures 1.19–1.21). How would you expect the mechanism of feeding and digestion in annelids to differ from that of platyhelminths and nematodes?

◆ The presence of a layer of muscle around the gut in annelids would enable them to move the guts independently of the rest of the body, which may facilitate eating larger prey. The body cavity of nematodes is not divided by septa, and the outer layer of the body wall secretes a tough, non-living layer of **cuticle** over the whole surface of the body. Thus encased, the nematode body cannot undergo changes in shape as large as those seen in annelids and platyhelminths. Like the flatworms, nematodes have a muscular pharynx (Figure A.24), but the posterior parts of their guts lack muscles. Food is taken in through the pharynx, and movements of the whole worm cause it to pass along the gut.

This muscular structure is unique to nematodes. Muscle fibres form in large cells, other parts of which have no contractile activity at all. Nematode cuticle is a tough, flexible but relatively inextensible material. Roundworms move by a thrashing motion that, although it is fast and appears to be powerful, is not very effective for burrowing or crawling on land.

1.6.4 Arthropods

The animals shown in Figure 1.23 and Plate 11 all belong to the largest phylum in the Animal Kingdom, the Arthropoda ('jointed feet'). Figures 1.24a and b show a side view and cross-section of a typical arthropod.

◇ Which of the phyla described in Sections 1.6.1–1.6.3 does the basic body plan of arthropods most closely resemble?

◆ The annelid body plan; as in annelids, much of the arthropod body is divided into segments and the gut has an anus as well as a mouth. The nervous system consists of two parallel nerve cords and there are muscles around the gut.

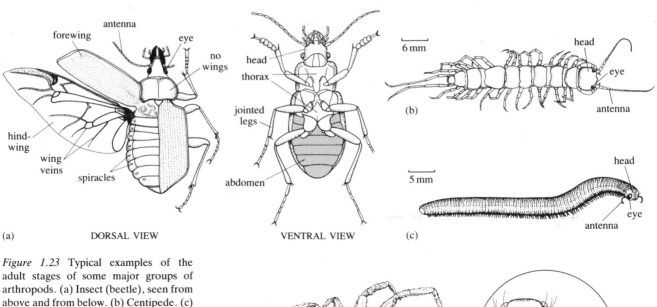

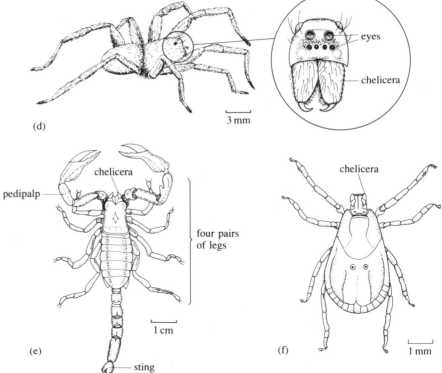

Figure 1.23 Typical examples of the adult stages of some major groups of arthropods. (a) Insect (beetle), seen from above and from below. (b) Centipede. (c) Millipede. (d) Spider. (e) Scorpion. (f) Mite. Note the different scales.

However, in contrast to annelids, there is a stiff outer 'shell' called the **exoskeleton** around the body, and pairs of jointed limbs.* The exoskeleton is made of cuticle, which consists almost entirely of proteins and carbohydrates in most terrestrial and freshwater arthropods, but in marine species such as crabs, the organic material may be impregnated with inorganic salts such as calcium carbonate.

*A limb may be defined as an appendage into which viscera such as guts and reproductive organs do not extend. The most usual function of limbs is in locomotion and feeding, but they can also, as in shrimps, act as respiratory organs (see Chapter 2, Section 2.3.2 and the Appendix).

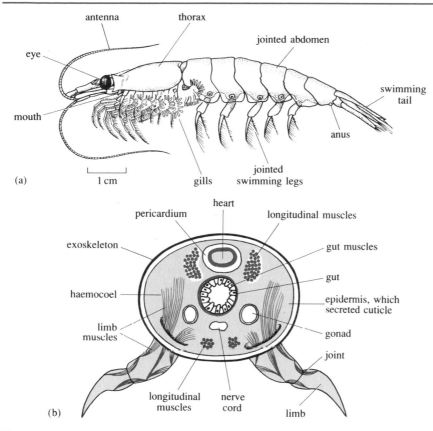

Figure 1.24 The basic structure of an arthropod. (a) A free swimming crustacean. The thoracic and abdominal appendages of the left side only are shown. (b) A cross-section through the middle of the body of an adult crustacean to show exoskeleton, body muscles (red) and haemocoel (pale grey).

The presence of a hard outer covering also means that most of the body surface is impermeable to gases. All but the very smallest arthropods have specialized respiratory organs; the two commonest types are protruding, frill-like **gills** in aquatic forms, and, in terrestrial arthropods, **tracheae** (sing. trachea). Tracheae are air-filled tubes that extend from openings at the surface, called **spiracles** (Plate 11a) through the body wall and into the muscles and other organs. In some larger arthropods, the spiracles have flaps and can be closed when the animal is resting or is exposed to toxic gases. Many arthropods, particularly small ones such as insects, do not have oxygen-carrying pigments in their blood, which therefore appears colourless. The coelom is greatly reduced in arthropods, but the blood vessels expand to form a **haemocoel** which assumes many of the functions of the coelom in other phyla. Thus the blood circulates partly through closed vessels, including the heart, and partly through the haemocoel.

Arthropods grow and change in shape by **moulting**: from time to time, the old exoskeleton is shed, revealing newly formed, soft cuticle that may take from a few minutes to a few days (depending upon the size of the animal) to harden. The body expands (by taking in water or air) while the new skeleton is still soft and extensible, so the new exoskeleton 'sets' at a larger size, and often in a different form. Arthropods do not feed while moulting and are often at greatest risk from predators while their protective cuticle is still soft. Large species (e.g. lobsters) may undergo more than a dozen moults before reaching mature size. If a limb or antenna of an immature animal is lost or damaged it may regenerate partially in subsequent moults.

In insects, spiders and some shrimps and crabs, fertilization takes place internally and an impermeable shell of cuticle is secreted around the eggs before they are laid. Mating is often preceded by visual, auditory or chemical signals by which the sexes identify and locate each other. Flies and beetles copulating end to end on the ground (Plate 11i) and dragonflies that mate during flight are a common sight on warm days in the spring and summer. In some arthropods, the juveniles are similar to the adults in structure and habits (e.g. spiders, woodlice) but many others, particularly insects, have a larva that is very different from the adult in appearance, habitat and diet (Plate 11k). (Larvae are discussed in Chapter 2, Section 2.2.1.) In most arthropods, parental care is limited to laying the eggs in a place where they are likely to find suitable food, but a few, such as wasps, bees and scorpions, feed and protect their young. A few species of insects, notably tsetse flies (which transmit trypanosomes, the protistans that cause sleeping sickness in humans and cattle) and aphids, are **viviparous**: the egg hatches and the larva completes most of its development inside the female's body.

Arthropods are by far the most abundant animals on land: centipedes (Plate 11e and Figure 1.23b), millipedes (Plate 11f and Figure 1.23c), woodlice (Figure 2.15), insects (Figure 1.23a), spiders (Figure 1.23d and Plate 11c), mites (Figure 1.23f), ticks (Plate 11b), scorpions (Figure 1.23e) and several less familiar groups live mainly on land. Of these the class Insecta (insects) is the most diverse (Plates 7c and 11a and 11i–n). The body of adult insects (Figure 1.23a) is clearly divided into three parts, the head, thorax and abdomen; typically there are three pairs of legs, a pair of antennae, several grasping, biting or sucking appendages around the mouth and, in the adults of many species, two pairs of wings. Juvenile insects never have functional wings. Many insect larvae have three pairs of legs similar to those of adults, but some have more than three pairs of leg-like appendages (e.g. caterpillars) and some are legless (maggots). Most insects are terrestrial throughout their lives, although the larvae of several large groups, including dragonflies, mayflies, caddis-flies and mosquitoes, live in freshwater, and a few, notably some beetles and bugs, are also aquatic as adults.

The most abundant kinds of insects are the beetles (order Coleoptera) (Plates 11a, m and n) and bugs (orders Hemiptera and Homoptera). The Diptera (two-winged flies) (Plate 11i) are among the most familiar because, in spite of strenuous efforts to eradicate them, several kinds continue to impinge upon human lives. Houseflies (*Musca domestica*), bluebottles, mosquitoes and tsetse flies are all dipterans, as is the fruit-fly *Drosophila* (Figure 1.25), the organism in which many of the basic principles of genetics were discovered. At summer temperatures, the life cycle of most species of *Drosophila* is complete in a few weeks: the females are attracted to ripe fruit or rotting vegetation on which they lay their eggs. After a few days, the legless, maggot-like larvae hatch and eat the fruit and any yeast or bacteria growing on it. After several larval stages and a brief pupal stage, the winged adults emerge, fly to new food sources, mate, lay eggs and the life cycle starts again. *Drosophila* owes its position as one of the most thoroughly studied of all organisms (with the possible exception of humans and *E. coli*) to its short life cycle and straightforward nutritional needs: the technology for breeding fruit-flies in milk bottles on a diet of yeast extract and molasses was perfected in the first decade of this century. They are now used in laboratories all over the world for studying basic mechanisms in genetics, developmental biology, neurobiology and many other topics.

Spiders, mites, ticks, scorpions and some minor groups are classified together as Arachnida (Figure 1.23d, e and f). Almost all of them are terrestrial, although a very few species spend part of their lives in freshwater. There is no distinct head or neck in arachnids; the body of spiders and scorpions is divided into two distinct regions and that of mites, harvestmen and other small arachnids forms a single, non-articulated mass. There are normally four pairs of walking legs (three in very small species) and two pairs of grasping

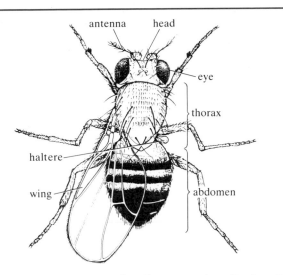

Figure 1.25 The fruit-fly, *Drosophila melanogaster* (class Insecta, order Diptera), one of the most thoroughly studied of all organisms. The right wing has been removed to show the haltere on the posterior thoracic segment. Halteres are sensory structures that act like a gyroscope, and evolved from reduced and modified hind wings.

appendages in front of the mouth (chelicerae and pedipalps, Figure 1.23e). Arachnids never have wings or true antennae. Most arachnids are carnivores, preying mainly upon other arthropods, particularly insects; the spider's web is by far the most successful means of catching flying or walking insects yet evolved. Most mites are tiny (0.2–2.0 mm long) and live in the soil or as parasites on plants or animals, including other arthropods. The time between fertilization and maturity can be as little as a week, and under favourable conditions they multiply very fast. Ticks (Plate 11b) are exceptionally large mites (up to 3 mm long) many of which feed by sucking blood from reptiles, birds or mammals.

The principal group of aquatic arthropods is the Crustacea, which includes crabs (Plate 11g), shrimps, prawns, lobsters, barnacles (Plate 11h) and krill. Some marine crustaceans, particularly copepods (Figures A.32 and A.33) such as *Calanus* are enormously abundant in the plankton of arctic and temperate oceans, where they feed on unicellular algae and other microorganisms and are themselves eaten by fish and whales. Crustaceans are also abundant in freshwater, including crayfish and many different kinds of shrimp-like arthropods such as water fleas (Figure A.31a). A few groups of crustaceans spend their entire lives on land, albeit in damp habitats. The most abundant and familiar are the Isopoda (woodlice; see Figure 2.15), but there are also crabs living in inland tropical rain forests, including a large, scarlet species in the cloud forest on mountains in central Africa. Fully aquatic crustaceans respire through gills attached to the legs and do not have tracheae. The distance between the respiratory surface where oxygen is absorbed and the tissues that use it is greater in such arthropods, and the blood system and heart of crustaceans, particularly in large species, are often more elaborate than those of insects. Many crustaceans have oxygen-carrying blood pigments that consist of protein and copper ions, which are colourless or slightly bluish in solution. Some of the largest living arthropods are crustaceans; some giant crabs can grow to a leg span of more than 2 m and some extinct aquatic arthropods were even bigger.

1.6.5 Molluscs

Snails, slugs, oysters and octopus belong to the second largest phylum in the animal kingdom, the Mollusca ('soft'). Figure 1.26a is a diagram showing body structure and Figure 1.26b a cross-section through a periwinkle, a snail-like mollusc that is very common on rocky beaches in southern Britain and elsewhere.

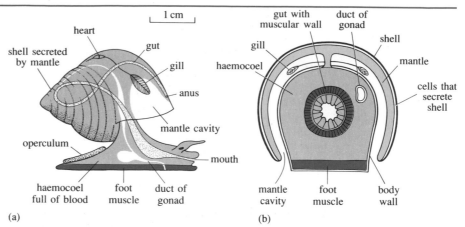

Figure 1.26 The structure of a typical gastropod mollusc, such as the periwinkle, *Littorina*. (a) Vertical section to show the arrangement of the gut (grey stipple), gonad (white). (b) A transverse section across the body to show the principal muscles, the mantle (pink) and the haemocoel (pale grey).

◇ List some similarities and differences between the body plan of this mollusc (see Figures A.25 and A.26) and those of nematodes (Figure 1.20), annelids (Figures 1.21 and 1.22) and arthropods (Figures 1.23 and 1.24).

◆ The molluscan gut resembles the guts of all these three phyla in having an anus as well as a mouth, and it is surrounded by muscles, like that of annelids and arthropods. As in arthropods, the coelom is greatly reduced and the body cavity is a haemocoel; there are gills. In contrast to all these phyla, there are no limbs and the body is not divided into segments.

The molluscan body plan has many unusual features: the gut, heart and gonads are coiled into a 'visceral mass' situated on top of the muscular 'foot'. The anterior part of the foot is called the head because the mouth, tentacles, eyes and much of the nervous system are located there. The visceral mass is covered with a flap of tissue called, quite logically, the **mantle**. The outer surface of the mantle secretes the hard shell, and the space between it and the body wall is open to the outside world. In many molluscs, gills protrude into the space between the body and the mantle, but in others such as the common or garden snail, the lining of the mantle acts as a respiratory surface. The mantle cavity is open when the animal is active, but the space is occluded when the snail withdraws its foot and head into the shell, using muscles that extend from the shell down into the protrusible parts of the body.

◇ If you can, watch a pond or garden snail or a slug moving along the wall of an aquarium or across a glass plate. What does the movement entail?

◆ The mollusc secretes a stream of sticky mucus at its anterior end, and propels itself through the trail so formed by waves of contraction of the muscles in the sole of the foot. In slugs, the shell is reduced or absent but the mantle is still present as an oval flap of tissue with an opening on the right of the body. You can observe these features very easily in any garden slug that is actively feeding or crawling.(Plate 13d).

Molluscan crawling is therefore similar to the movement of soft-bodied worms such as free-living platyhelminths (see Section 1.6.3). Locomotion can also be assisted by changes in body shape, for which the fluid in the haemocoel acts as a hydrostatic skeleton in much the same way as does the coelomic fluid of worms.

Snails (Plates 13b and c) and slugs (Plate 13d) belong to the largest class of molluscs, the class Gastropoda ('stomach-foot'; so called because part of the gut is in the locomotory foot) which also includes numerous marine,

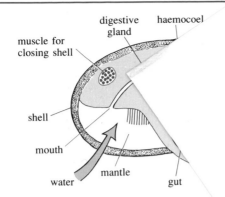

muscle for
closing shell

digestive
gland

haemocoel

le for
shell

shell

mouth

water

mantle

gut

Figure 1.27 A typical bivalve mollusc, with one shell valve removed, showing the muscles (red) and the haemocoel (pale grey). The pink arrows show the direction of the water currents over the gills.

freshwater and terrestrial species, including s   serts, the deep sea and other inhospitable places.   c features include a shell, which may be a simple cone (a   ) or elaborately coiled (as in periwinkles and garden snails) or reduced to a sliver (as in slugs), and paired tentacles on which there are simple eyes and other sense organs. In most gastropods, the dorsal side of the body containing the viscera and the mantle is twisted round relative to the head and foot (Plate 13b). This rather odd phenomenon is known as torsion, and results in the gills and the opening of the mantle cavity being over the anterior end of the body, near the head (Figure 1.26a). Many gastropods are herbivores, scraping off encrusting algae and the surface of angiosperm plants with rhythmic movements of the horny, rasp-like structure called the **radula**. The radula forms on a muscular structure in the mouth; repeated scraping wears down the elaborately patterned abrasive surface, but as the radula grows continuously, its worn surface is replaced.

◇ List some features of mussels (Figure 1.27 and Figure A.27) and squids (Figure 1.28) that show that they too should be classified as molluscs.

◆ Like gastropods, these animals have a mantle and a muscular foot closely associated with the mouth and nervous system. There is no segmentation and the gut has two openings.

These features are sufficient grounds on which to classify these animals as molluscs, but they also have some fundamental differences from gastropods and from each other. Mussels, cockles, oysters, scallops and similar animals are bivalves (also called lamellibranchs and pelecypods in some older texts). As the name implies, the bivalve shell is in two (usually symmetrical) parts, joined by an articulating hinge. Both the mantle and the shell are relatively large and enclose the entire body. The foot is relatively small (in some forms such as oysters, it has almost disappeared or has become non-muscular), and much of its volume is occupied by the gut, which is relatively massive although the mouth is small and there is no sign of hard parts around it. The gills, however, are large and occupy much of the mantle cavity. Most bivalves are sessile, and feed on detritus and very small animals and plants which they extract from water currents that pass over the gills. In bivalves that live in crevices or buried in sand, water is pumped into the gill chambers by muscles or cilia, and small food particles are trapped in mucus that is secreted by and suspended between the elaborately frilled gills. Bivalves do not have a radula or beak, and never scrape or bite their food. Some bivalves such as mussels attach themselves to surfaces exposed to water currents, such as rocks, driftwood, and other animals including large fish and whales. Although bivalves are common on intertidal rocks and beaches, only a few species live in freshwater, and none is truly terrestrial.

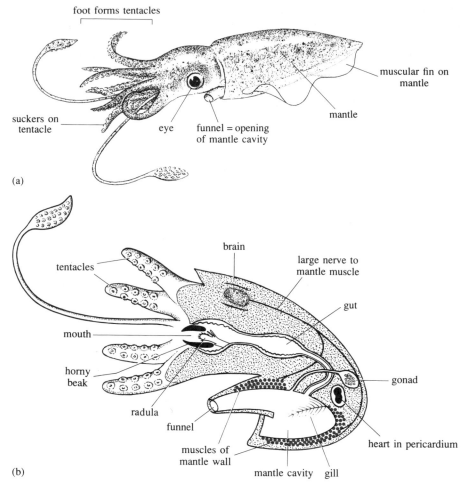

foot forms tentacles

muscular fin on mantle

mantle

suckers on tentacle

eye

funnel = opening of mantle cavity

(a)

brain

large nerve to mantle muscle

tentacles

gut

mouth

horny beak

gonad

radula

funnel

muscles of mantle wall

mantle cavity

gill

heart in pericardium

(b)

Figure 1.28 A typical cephalopod mollusc (squid *Loligo*). (a) Whole animal. Fast, powerful extension of the two long, thin tentacles stuns prey such as prawns and small fish. The prey adheres to the pads of suckers and is pulled towards the mouth, where it is held by the other tentacles and 'chewed' by the beak and radula. (b) Longitudinal section to show the arrangement of the principal organs. The brain is partially enclosed in a cartilaginous 'skull'.

Squids, octopus and cuttlefish are cephalopods ('head-foot'). The mantle is thick and muscular and, except in a few atypical forms, the shell has almost disappeared.

◇ In Figure 1.28, does the mantle and the mantle cavity of the squid retain some of the same functions that it has in bivalves and gastropods?

◆ Yes. As in other molluscs, the gills in the mantle cavity are respiratory organs.

Cephalopods swim slowly by undulating movements of the fins, formed from muscular extensions of the mantle. They swim fast by a form of jet propulsion in which the muscles of the mantle cavity cause it to expand, sucking in water, which is then squirted out through the funnel. They normally swim backwards, with the tentacles trailing, but they can move in any direction, depending upon which way the funnel is pointing. Such powerful movements involve fast, well coordinated contraction of the muscles, and cephalopods have a large, powerful heart and an elaborate nervous system.

◇ Compare the structure of the head of the periwinkle (Figure 1.26) with that of the squid (Figure 1.28).

◆ The squid's mouth is equipped with a stout beak and a radula; the nervous system is expanded to form a large brain and there are two large eyes.

All cephalopods are marine and the majority live in the open oceans, sometimes at great depth; most of them are active predators, chasing and killing fishes and large invertebrates. *Octopus* is among the most familiar cephalopods because it is exceptional in living in shallow, coastal waters.

1.6.6 Echinoderms

Starfish (Figure 1.29 and Figure A.38a) and sea-urchins (Plate 14a and Figure A.38c) are the most familiar members of the phylum Echinodermata ('spiny skin'). Their most striking feature is their **pentamerous** (fivefold) **symmetry**. Both the mouth and the anus are near the centre of the body, the former normally on the ventral surface and the latter on the dorsal surface. The 'arms' cannot strictly be called limbs because branches of the gut extend into

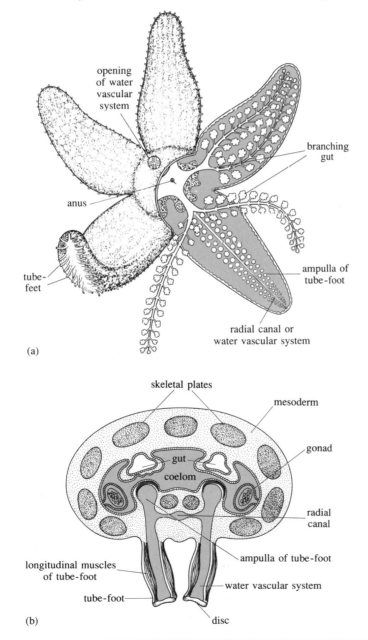

Figure 1.29 A typical echinoderm (starfish *Asterias*). (a) As seen from above, with two arms dissected to reveal parts of the gut, ampullae of tube-feet and radial canals. (b) A transverse section through an arm to show the muscles (red), skeleton (dark grey), gut (white), coelom and ampullae of tube-feet (grey).

them. They also have an internal skeleton and, in many species, spines that protrude through the soft tissues. The nervous system consists of a network of nerve cells concentrated around the mouth. There is no blood vascular system but echinoderms have several coelom-like cavities.

The **tube-feet** and the **water vascular system** (Figure 1.29) are unique to echinoderms. The structure of typical tube-feet is shown in Figure 1.29b. Each tube-foot consists of an internal ampulla and a portion that protrudes through the skeleton. The base is a disc, often in the form of a suction pad or sticky with mucus. The shaft of the tube-foot can be shortened by contraction of the muscles along its walls, and can be extended by forcing fluid through the water vascular system and from the ampulla into the shaft. The pressure inside individual tube-feet can be adjusted by flow of fluid along the narrow tubes that link rows of them together (Figures 1.29a and b). The muscles of the shaft can also make the tube-foot wave from side to side and many are equipped with sense organs sensitive to touch and to chemicals. Free-living echinoderms can 'creep' slowly across the substratum by means of waves of attachment and detachment of the tube-feet.

Echinoderms (Plate 14) have from dozens to many thousands of tube-feet which are involved in activities such as respiration and attachment to the substratum as well as in locomotion and feeding. For example, in some starfish (class Asteroidea), including the common British species that preys on mussels and other bivalves, the tube-feet can act as suckers, sticking the arms to the prey's shell. The arms fold around the prey, then gradually extend, prising the two valves apart and exposing the soft tissues inside.

Sea-urchins (Plate 14a), heart urchins (sea potatoes) and sand-dollars (Figure 2.4) are echinoids ('spiny'). They have long, muscular tube-feet, some of which can reach out beyond the spines. In many species, the spines can be moved independently by muscles at the base and they vary in shape from short and stubby, to long and razor-sharp, sometimes containing powerful toxins. The combination of long, powerful tube-feet and mobile spines enable these echinoderms to 'walk' or burrow in soft sand and over rocks and gravel. Some species can wedge themselves into crevices, often surprisingly tightly. Most echinoids live buried in sand or in sheltered waters on coral reefs or intertidal rocks, where they scrape off encrusting organisms, such as algae and corals, using the sharp teeth that form part of a unique, jaw-like structure called Aristotle's lantern (Plate 14a). Sand-dollars (see Chapter 2, Figure 2.4) are flattened echinoids with numerous, very small spines and many thousands of tiny tube-feet and a greatly reduced Aristotle's lantern. The guts are found to be full of sand, and when living sand-dollars are examined with a powerful microscope in the seawater in which they are found, tiny, food-gathering tube-feet can be seen to pick up diatoms (Section 1.3.2) one by one. The food particles are passed along the rows of tube-feet to the mouth, where they are crushed by the Aristotle's lantern, together with sand grains. Up to 3 per cent of the surface of a sand grain is covered with various kinds of prokaryotes (Section 1.2), sessile protistans and other encrusting organisms, which are digested along with the slurry of crushed diatoms.

The other major living classes of echinoderms are the brittle-stars (Figure A.38b) and basket stars (ophiuroids, 'snake-like'), which derive their name from their long, highly mobile arms (see Chapter 2, Figure 2.12); the sea-fans and feather stars (crinoids, 'crown-like'; Figure A.38e), and the elongated, worm-like holothurians (sea-cucumbers) in which the arms and spines are greatly reduced and the body is soft and extensible (Figure A.38d and Plate 14b). Some holothurians have the rather strange habit of eviscerating themselves when attacked by predators. In Plate 14c, you can see the guts containing sand from which minute organisms and detritus were being digested. Like most other echinoderms, these groups are most abundant and diverse on the sea floor, particularly in tropical oceans. Crinoids occur mainly in deep water, but holothurians and ophiuroids are found in the intertidal zone, in shallow seas and on the floor of deep oceans.

1.6.7 Chordates

Fishes, amphibians (frogs and newts), reptiles, birds and mammals, together with some less familiar forms, all belong to the phylum Chordata. Although they may be the most familiar kind of animals, they are by no means the most abundant; only about 5% of all animal species so far described are chordates, and half of these are fishes, most of them living in the sea. Figures 1.30a and 1.30b are, respectively, a cross-section and longitudinal section of a bony fish. Chordates have several features in common with other phyla, including muscles around the gut, a coelom and an internal skeleton.

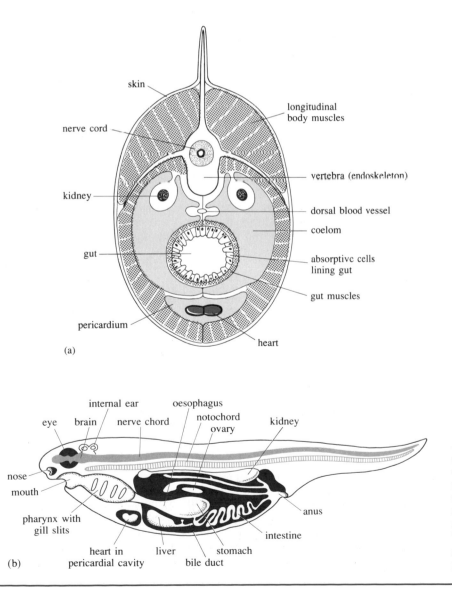

Figure 1.30 Diagrams to show the basic structure of vertebrates. (a) Cross-section of a fish to show the skeleton, body muscles (dotted red), heart (solid red) and coelom (grey). (b) Vertical section of a fish to show the nerve chord and notochord (grey) and the coelom (black). The mouth and the anterior part of the gut are cut open to show the gill slits that perforate the pharynx.

◇ List some differences between the chordate body plan and that of the other phyla already described.

◆ There is a stout skeletal rod surrounded by muscles near the centre of the body, to which the nerve cord is dorsal and the heart is ventral. The main coelom and gut are ventral and most of the coelom is in the middle of the body; it does not extend into the head or tail.

Although the muscles, spinal nerves and some of the blood vessels are arranged serially, there is no sign of partitioning of the abdominal coelom. There are layers of muscles around the gut and near the body wall. The stout skeletal rod supporting the serially arranged muscles is the **notochord** from which the phylum Chordata takes its name. The gut and body wall are perforated at the pharynx, the region of the gut just behind the mouth, to form a series of slits through which fluids taken in through the mouth can be evacuated before they reach the stomach. In aquatic chordates, these pharyngeal slits support the gills, which are the principal respiratory organs and sometimes play a part in feeding.

Chordates are divided into the invertebrate chordates, the Urochordata ('tail-cord'; sea-squirts, salps and their relatives) and Cephalochordata ('head-cord'; small, marine, fish-shaped chordates, of which the best-known is the amphioxus) and the much more abundant and familiar Craniata, also known as Vertebrata. The names refer to the cranium, a hard case around the brain, and to the vertebrae, which are serially arranged skeletal elements that encase the nerve cord and, in most cases, the notochord. In almost all living vertebrates, the notochord is conspicuous only in the early embryo and is replaced by vertebrae before birth or hatching.

Most urochordates and cephalochordates are filter feeders with a greatly enlarged pharynx and numerous gill-slits. As well as acting as respiratory surfaces, the gills trap minute food in the cilia and mucus on their surface. There is no hard skeleton in the invertebrate chordates; the principal skeletal element is the tough, flexible notochord that contains both muscle and a deformable jelly-like material that bends and shortens actively during swimming. There is more about the structure and life cycles of chordates in Chapter 2, Section 2.2.1.

Vertebrates

The vertebrate chordates include mammals, birds, reptiles, amphibians, and several different groups of fishes. The skeleton is much more extensive than the basic plan shown in Figure 1.30 and in the great majority of species, it consists of a hard tissue, bone, that is found only in vertebrates. Bone consists of an extracellular matrix of a fibrous protein, collagen, that is impregnated with minerals, mostly calcium phosphate. The **cranium** and vertebrae form a hard case around the brain and spinal cord, replacing much of the notochord. An upper and a lower jaw, often bearing teeth, may be attached to the cranium, these together forming the skull, and skeletal elements involved in respiration (gill bars in most fishes and larval amphibians and ribs in terrestrial vertebrates) are attached via muscles to the vertebrae. The vertebral column extends beyond the anus, forming the **tail**, which is the principal locomotory organ in most aquatic forms. There are also two girdles, an anterior pectoral girdle (that forms the shoulder in mammals) and a posterior pelvic girdle (that forms the hip in mammals) which support paired fins or limbs. All these elements of the skeleton have many large muscles attached to them. The arrangement and properties of the organs of the vertebrate body are described in much greater detail in later chapters, so we give only a very brief outline of the main features here.

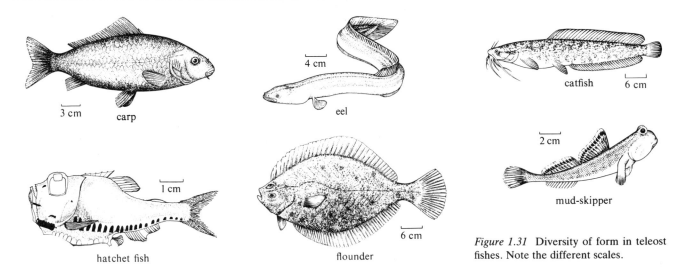

Figure 1.31 Diversity of form in teleost fishes. Note the different scales.

Fishes

The animals known collectively as 'fishes' belong to at least eight distinct groups, some of which are completely extinct. Most living fishes belong to the class Osteichthyes ('bony fish') of which by far the most abundant group are the teleosts ('complete bones'; Figure 1.31 and Plates 15b and c), which occur in freshwaters, shallow seas and the open ocean in a huge variety of shapes and sizes that reflect a wide range of diets and styles of swimming. Most food fishes, including trout and cod, and most aquarium fishes, including guppies and goldfish, are teleosts. In typical teleosts, the skeleton, fin rays and the scales contain bone, and the gill slits are covered by an articulated flap of bone, called the operculum (see Figure A.42c and Plate 15b). Most teleosts also have a gas-filled sac called the **swimbladder** in the dorsal wall of the abdomen (Figure 1.32) that reduces the density of the fish. Many teleosts actively adjust the volume of their swimbladder, and so can be neutrally buoyant at different depths.

The next most abundant group of living fishes is the Chondrichthyes ('cartilage fish'; Figure 1.33 and Plate 15a), which includes the sharks and rays, almost all of which are medium-sized or large oceanic fishes. As their name implies, the skeleton of chondrichthyans is composed entirely of cartilage, and the skin is tough and flexible rather than scaly. They do not have a gas-filled swimbladder and so are usually more dense than water. Chondrichthyans that do not live on the sea-bed maintain their depth in the water by swimming continuously. There is no operculum so the gill slits, normally five in number, are visible in the pharyngeal region behind the head (see Figures 1.33b and d).

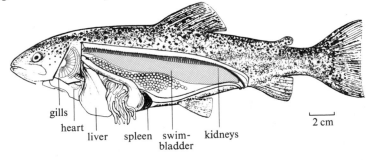

Figure 1.32 A rainbow trout dissected to display the internal organs. The five pairs of gills are normally covered by an operculum, which also acts as a pump, drawing water in through the mouth and out over the gills.

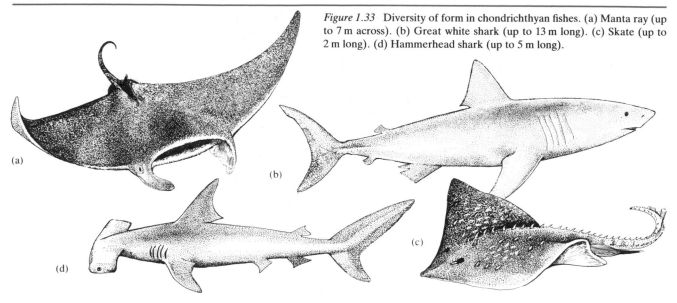

Figure 1.33 Diversity of form in chondrichthyan fishes. (a) Manta ray (up to 7 m across). (b) Great white shark (up to 13 m long). (c) Skate (up to 2 m long). (d) Hammerhead shark (up to 5 m long).

Tetrapod vertebrates

Amphibians, reptiles, birds and mammals are known collectively as **tetrapods** ('four feet'), even though some, such as snakes, do not have feet at all, and others such as birds and humans, have only two feet.

The most fish-like of the tetrapod vertebrates are the class Amphibia ('both lives'), which differ from fishes mainly in that the principal means of locomotion in adults is the legs (Plates 15d and e). As their name implies, most amphibians spend part of their lives in water and part on land; there is nearly always an aquatic larval stage—in the case of frogs, the familiar tadpole (Figures 1.34b and c; also see Chapter 2, Section 2.2.1). The principal living groups of amphibians are the Anura ('tailless'; the frogs and toads; Plate 15e) and the Urodela ('tailed'; newts; Plate 15d), of which the former are much more abundant and diverse, living in or near freshwater, or at least in damp places, on all continents except Antarctica. As adults, most anurans are active carnivores, eating insects, slugs and many other garden pests. One species of frog lives in tropical salt marshes, but no living amphibians are truly marine. Almost all invertebrates, fishes and amphibians are **poikilothermic**; their body temperature is always close to that of their surroundings.

Figure 1.34 Stages in the life cycle of a frog (order Anura). (a) Frogspawn. (b) Early tadpole larva. (c) Later tadpole with developing hind legs. (d) Adult frog. Not drawn to the same scale.

Frogs, particularly those of the genus *Rana*, have long been favourite subjects for experimental physiologists, because, living as they do in ponds of northern Europe and America, they remain active over a wide range of temperatures. Almost all their physiological processes remain viable at 0 °C, which is

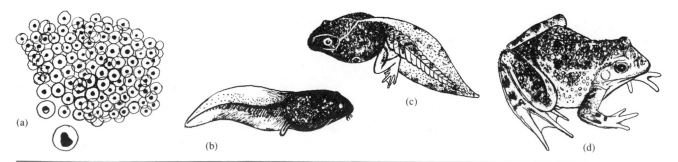

technically much easier to maintain in the laboratory than the higher temperatures required by mammals or birds. The frog can withstand higher temperatures for short periods of time, so its tissues continue to function in climates that are comfortable and convenient for physiologists.

The body plans and many of the physiological features of the class Reptilia (reptiles) (Plates 11h and 15f–i) and the class Aves (birds) (Figure 1.35 and Plate 17) are so similar that they could be classified together: both have an outer covering of impermeable horny scales, which in birds is modified to form feathers over most of the body. Both groups are thoroughly adapted to terrestrial life and many forms can survive long periods without freshwater. All birds and the great majority of reptiles reproduce by laying shelled eggs, called **amniote eggs** (Chapter 2, Figure 2.20), which are always laid on land. Reptiles and birds (and mammals) never have a larval stage. An important difference between reptiles and birds is that parental care is usually minimal in the former, but the latter nearly always brood their eggs and, in most cases, feed and care for the young after hatching. In modern birds the teeth are always replaced by a horny beak, but most reptiles (except turtles) have numerous teeth. Most reptiles are poikilothermic: their body temperature is variable, and rarely more than a few degrees different from that of the surroundings, because they warm themselves mainly by basking in the sun or some other external source of heat. Birds are **homoiothermic**; that is, most of their body heat is generated internally and they maintain a constant temperature of about 40 °C. The body is insulated by a layer of feathers, which also play essential roles in flight and the elaborate social, sexual and parental behaviour of birds.

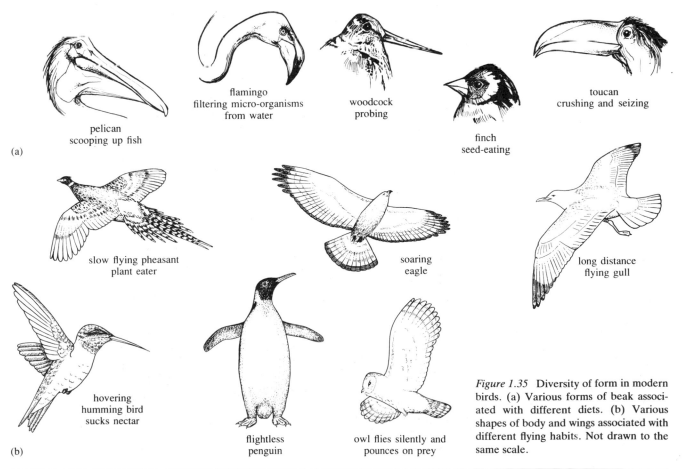

pelican
scooping up fish

flamingo
filtering micro-organisms
from water

woodcock
probing

finch
seed-eating

toucan
crushing and seizing

(a)

slow flying pheasant
plant eater

soaring
eagle

long distance
flying gull

hovering
humming bird
sucks nectar

flightless
penguin

owl flies silently and
pounces on prey

(b)

Figure 1.35 Diversity of form in modern birds. (a) Various forms of beak associated with different diets. (b) Various shapes of body and wings associated with different flying habits. Not drawn to the same scale.

Reptilian skin normally contains hard scales which, in the chelonians (tortoises and turtles), form a 'shell' covering most of the body. Reptiles were much more abundant and diverse in the past than they are now (see Chapter 2, Section 2.2.2). The principal living groups are the squamates (lizards and snakes; Plates 15h and 15i), the crocodilians (Plate 15f) and the chelonians (Plates 11h and 15g), which differ greatly from each other in the structure of the head, trunk and limbs. Although mostly terrestrial or freshwater, a few species of each group spend a substantial part of their lives in the sea, where they breathe air, and marine crocodiles and turtles all lay their eggs on the shore above the tide line. The majority of living reptiles (including all snakes, but excluding some tortoises and a few lizards) and many birds feed on other vertebrates or on large invertebrates such as worms, molluscs and arthropods.

There are more than twenty orders of birds, of which by far the most abundant and diverse are the passerines (song-birds: sparrows, robins, finches, crows, etc.). However, the basic anatomy and physiology are similar in all living species of birds and the orders are distinguished mainly by relatively minor differences in the structure of the beak and feet, many of which correlate with diet and habits, as shown in Figure 1.35. All birds, including those that feed mainly or entirely in the sea, build their nests and raise their young on land, and often migrate long distances to their breeding sites.

The class Mammalia (Plate 16) takes its name from the **mammae**, which are glands, normally on the ventral surface of the body that, in the adult female only, secrete milk that nourishes the young after birth (Plate 16b). Mammals are also homoiothermic, but their anatomy and reproductive system (Plate 16a) differ so fundamentally from those of birds that a constant body temperature, and with it the capacity for prolonged, strenuous activity, probably evolved independently in the two groups. Mammals always have hair in some form, normally as fur over most of the body, but sometimes as horny scales, spines, bristles, whiskers or eyelashes. Nearly all mammals have teeth, which may be large and complex in internal structure. However, in contrast to large reptiles such as crocodiles, in which worn or broken teeth may be replaced many times, in most mammals the dentition is replaced only once in a lifetime.

There are three living subclasses of mammals, distinguished mainly by their reproductive mechanisms. Monotremes include only the duck-billed platypus and the spiny anteaters (Figure 1.36), which occur only in Australia and New Guinea. Monotremes lay eggs in a simple nest but the young suckle milk, secreted, often in impressive quantities, from the mother's mammary glands. Kangaroos, wallabies, opossums, koala bears, numbats and their relatives are metatherians (also called marsupials; Figure 1.37) and are also most abundant and diverse in Australia and New Guinea, but a few species are also found naturally in North, South and Central America. The young are very small and anatomically immature except for the forelegs, with which they climb from the birth canal to nipples, which are often inside a pouch on the mother's abdomen, where they may remain until more than a third grown.

Humans, most domestic livestock and many of the most familiar wild animals are eutherian mammals (also called placentals). They occur on all continents except the antarctic mainland. The earliest eutherian mammals were similar in structure and general habits to living moles and hedgehogs; from them evolved about 20 orders (Figure 1.38 and Plate 16), of which by far the most abundant are the rodents (rats, mice, voles, hamsters, squirrels, guinea-pigs and many others).

(a)

(b)

Figure 1.36 Prototherian (monotreme) mammals. (a) Duck-billed platypus (adult body mass 0.5–2 kg). (b) Spiny anteater (adult body mass 2.5–6 kg). Not drawn to the same scale.

Figure 1.37 Various metatherian (marsupial) mammals. Koala (adult body mass about 8 kg). Wombat (adult body mass 15–25 kg). Numbat (adult body mass less than 0.5 kg). Kangaroo (adult body mass up to 90 kg). Not drawn to the same scale.

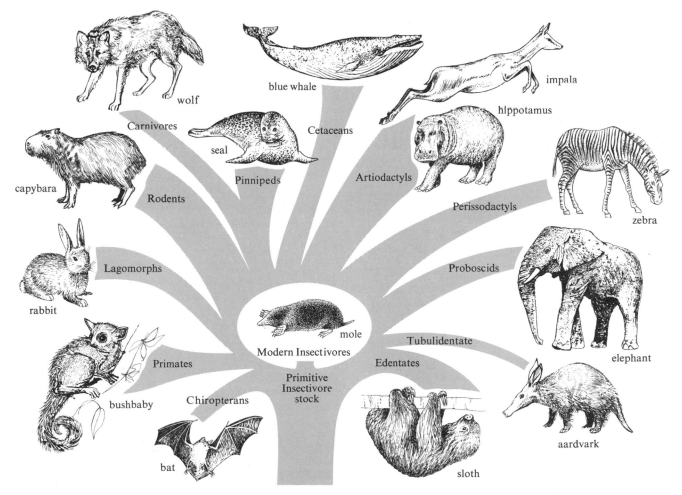

Figure 1.38 The principal groups of eutherian (placental) mammals. The earliest mammals were similar to modern insectivores and are believed to be the ancestors of all the other orders. Not drawn to the same scale.

◇ From Figure 1.38 and Plate 16, in which parts of the body are there the greatest differences between these orders of eutherian mammals?

◆ The proportions of the head and the limbs: all the legs of artiodactyls, perissodactyls and carnivores are elongated, but those of seals are reduced. In bats, the forelimbs are much larger than the legs, and in whales there are no external hind limbs and the forelimbs are very small.

One of the most significant features of mammals is that, unlike most other vertebrates, the majority of them are herbivores (e.g. kangaroos, koalas, almost all rodents, pigs, hippos, camels, deer, antelopes, horses, rhinoceroses, elephants, monkeys, apes, rabbits, hares).

◇ In which other groups are a high proportion of the species herbivorous?

◆ Insects, particularly beetles, bugs, butterflies and moths, and molluscs (e.g. slugs and snails).

Insects, mammals and gastropods (in that order) are now (and probably have been since the beginning of the Tertiary), the principal consumers of angiosperm and coniferous plants on land. Some of the implications of this situation for the anatomy and physiology of mammals, and their biological relations to other organisms, are discussed in the following chapters.

Summary of Section 1.6

The basic body plan is the most important criterion for the classification of multicellular animals. The body plan includes the arrangement of layers of living tissue and of the body cavities, the anatomical relations of the nervous system and muscles, and the composition and anatomical relations of the skeleton and locomotory appendages.

Table 1.3 Some features of the body plans of some animal phyla (for use with Question 7).

Phylum Class Common Name	Cnidaria		Platyhelminthes Turbellaria	Nematoda
	Medusa	Polyp	Planarian	Roundworm
Symmetry				
(radial (R), bilateral (B) or pentamerous (P))				
Cavities in the body				
Gut or enteron				
Pseudocoel				
Coelom				
Haemocoel or blood system				
Segmentation of all or part of the body				
Locomotory appendages (type)				
Skeleton (type)				
Anus present				
Musculature				
Muscles in gut wall				
Circular body muscles				
Longitudinal body muscles				
Muscles in locomotory appendage(s)				
Any special features not already mentioned				

Question 6 (*Objective 1.7*) Which of the statements (i)–(xii) describe typical features of the following phyla:

(a) Porifera	(d) Nematoda	(f) Mollusca	(h) Echinodermata
(b) Cnidaria	(e) Annelida	(g) Arthropoda	(i) Chordata
(c) Platyhelminthes			

   (i) Discrete body layers cannot be distinguished.

  (ii) Most living cells occur in the endoderm or ectoderm.

 (iii) Ectoderm, endoderm and mesoderm are present.

 (iv) There is a mouth but no anus.

  (v) There is both a mouth and an anus.

 (vi) There is neither mouth nor anus.

(vii) There is a hydrostatic skeleton (formed from coelom, haemocoel or pseudocoel).

(viii) There is a separate blood system as well as a body cavity (coelom, haemocoel or pseudocoel).

 (ix) The coelom is divided into segments.

  (x) There is an exoskeleton.

 (xi) There is an endoskeleton.

(xii) Tube-feet are present.

Question 7 (*Objectives 1.7 and 1.8*) Provide your own summary of the major anatomical features of the animal phyla described in Section 1.6 by completing Table 1.3 with + signs to indicate the presence of features and − signs to indicate their absence, or write in the relevant words. The Table will be a useful reference when you are reading later chapters.

Annelida	Arthropoda	Mollusca		Echinodermata	Chordata
Polychaeta	Crustacea	Gastropoda	Cephalopoda		
Ragworm	Shrimp	Snail	Squid	Starfish	Fish

Question 8 (*Objective 1.8*) Classify the following organisms as precisely as you can:

(a) The animal is bilaterally symmetrical. The whole nerve cord is dorsal to the gut. The skeleton is internal and it and the muscles extend beyond the anus, forming a tail. Lungs and teeth are present and the skin contains hard scales.

(b) The bilaterally symmetrical, segmented body is divided into a head, thorax and abdomen. The animal has a tough outer cuticle and pairs of openings leading to tubes containing air.

(c) The animal is flattened and elongated and consists of serially arranged sections, growing from a 'head'. It is an internal parasite.

(d) The animal has a muscular foot that secretes mucus, a head with two small eyes and two tentacles, and it respires through the wall of a cavity formed from a muscular flap of tissue over the body.

(e) The animal is an elongated, soft-bodied, bilaterally symmetrical worm with deformable segments and without paired limbs or muscular projections.

OBJECTIVES FOR CHAPTER 1

When you have completed this chapter, you should be able to:

1.1 Define and use, or recognize, definitions and applications of each of the terms printed in **bold** in the text and the principal taxonomic terms.

1.2 Describe the structure and properties of typical viruses. (*Question 1*)

1.3 Describe and recognize the differences between prokaryotes and eukaryotes. (*Question 2*)

1.4 Describe the structure and properties of bacteria and cyanobacteria. (*Question 3*)

1.5 Outline the criteria that define the major taxonomic divisions of the plant kingdom. (*Question 4*)

1.6 Describe the basic structure and properties of fungi. (*Question 5*)

1.7 List and recognize the essential features of the body plans of the following phyla: Porifera, Cnidaria, Platyhelminthes, Annelida, Nematoda, Arthropoda, Mollusca, Echinodermata, Chordata. (*Questions 6 and 7*)

1.8 Name and describe some typical features of the major living classes of the following phyla: Platyhelminthes, Annelida, Arthropoda, Mollusca, Chordata. (*Questions 7 and 8*)

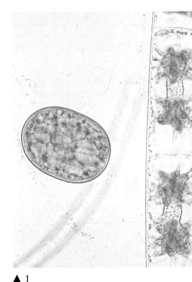

▲ 1

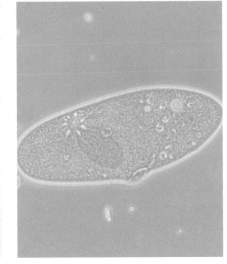

▲ 2 (a)

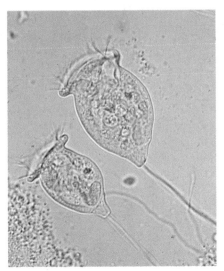

▲ 2 (b)

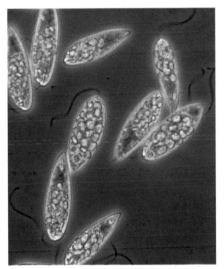

▲ 3 (a)

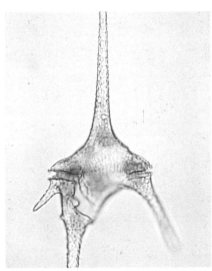

▲ 3 (b)

3 (c) ▼

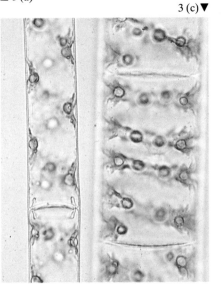

3 (d) ▼

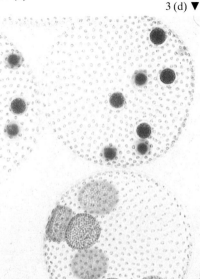

Plate 1 Left: an exceptionally large pro-karyotic cyanobacterium (diameter 50 µm). *Right:* a filamentous eukaryotic green alga (class Chlorophyceae, *Zygnema* sp.). Note the nucleus and chloroplasts in the eukaryo-tic cells, and the lack of membrane-bound structures in the prokaryote.

Plate 2 Protistans.

(a) *Paramecium* (250 µm), class Ciliata. Note the cilia, nucleus and 'gullet' and food vacuoles.

(b) *Vorticella* (60 µm), class Ciliata. A stalked form with spirally arranged cilia, 'gullet' and large nucleus.

Plate 3 Unicellular algae.

(a) *Euglena* (50 µm), a solitary freshwater alga (division Euglenophyta). Note the long flagellum and numerous chloroplasts.

(b) *Ceratium* (300 µm), a dinoflagellate (di-vision Pyrrophyta). Cellulose plates form the three spines and there is a flagellum in the groove around the middle.

(c) *Spirogyra* (40 µm and 100 µm), a fresh-water filamentous alga (division Chloro-phyta, class Chlorophyceae). The spirally arranged chloroplasts are typical.

(d) *Volvox* (1 mm), a colonial green alga (division Chlorophyta, class Chloro-phyceae). Daughter colonies are forming inside the main colony.

▲ 3 (e)

▲ 4

▲ 5 (a)

(e) A lichen on a tree in winter. Lichens are an intimate mutualism between an alga or a cyanobacterium and a fungus (usually an ascomycete).

Plate 4 Multicellular algae. Bladder wrack (division Phaeophyta, *Fucus vesiculosus*). A brown alga that is often massive and abundant on shores and in tidal pools. Note the flat thallus and air-bladders.

Plate 5 Division Chlorophyta, class Bryophyta.

(a) *Marchantia* sp., a liverwort growing among leaf litter. Most of the tissue is the gametophyte, and the umbrella-like structures are sporophytes.

(b) *Tortula muralis*, a moss growing among rocks. The gametophyte forms a cushion-shaped structure and the sporophytes are tall with slender capsules. Although the gametophyte can survive drought conditions, water is essential for completion of the life cycle.

Plate 6 Tracheophyte plants that do not bear seeds.

(a) Class Lycopsida. *Lycopodium clavatum*, a club moss growing among grasses. The leafy shoots arise from a rhizome and carry reproductive structures (sporangia) at the bases of the leaves near the tip of the shoot.

(b) Class Sphenopsida. *Equisetum* sp., a horsetail growing on moorland. The leaves are arranged in whorls on all shoots and a central thick shoot bears the reproductive structures.

(c) Class Pteropsida, subclass Filicidae. A tree fern growing in dense rainforest in the mountains of East Africa. Most temperate zone ferns are much smaller.

▲ 5 (b)

▲ 6 (a)

6 (b) ▼

6 (c) ▼

▲ 7 (a)

▲ 7 (b)

▲ 7 (c)

▲ 7 (d)

▲ 7 (e)

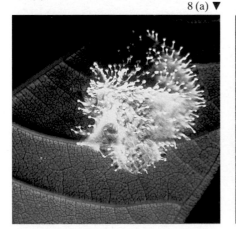

8 (a) ▼

8 (b) ▼

Plate 7 Seed-bearing tracheophyte plants.

(a) A branch of a larch tree (class Pteropsida, subclass Gymnospermidae, family Pinaceae, *Larix* sp.). These old female cones have opened, exposing the seeds.

(b)–(e) Class Pteropsida, subclass Angiospermidae (flowering plants).

(b) A branch of the birch (superorder Dicotyledonae, family Betulaceae, *Betula* sp.) a wind-pollinated tree. The larger male flowers form hanging catkins and the smaller female flowers are upright spikes.

(c) A butterfly (class Insecta, order Lepidoptera) sips nectar from the flowers of the blackberry (superorder Dicotyledonae, family Rosaceae, *Rubus* sp.) in which numerous stamens are visible.

(d) Flowers of the white deadnettle (superorder Dicotyledonae, family Labiatae, *Lamium album*) are pollinated by bees. The stamens arch under the large upper petal and brush against the bee when it alights on the lip formed by the lower petals.

(e) A bee orchid (superorder Monocotyledonae, family Orchidae, *Ophrys apifera*). This flower resembles the abdomen of a female bee so closely that male bees attempt to copulate with it, thereby brushing pollen onto themselves and transferring it to the next flower that they visit.

Plate 8 Fungi.

(a) A primitive fungus in which the mycelia grow over and into a dead insect, feeding on its tissues. Spores form on the projections.

(b) A mushroom, the reproductive structure of a basidiomycete. Most of the mycelia are underground, feeding on decaying plant and animal tissues.

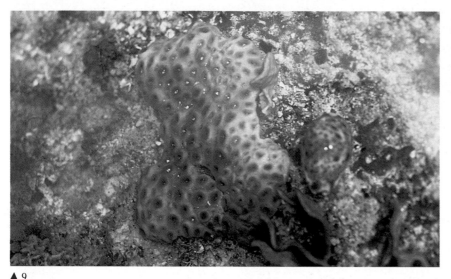

▲ 9

▲ 10 (a)

▲ 10 (b)

▲ 10 (c)

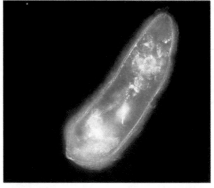

▲ 10 (d)

10 (e) ▼

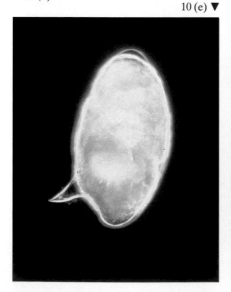

10 (f) ▼

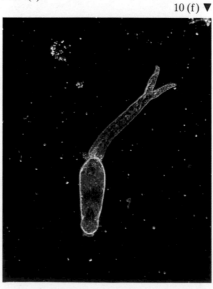

Plate 9 A sponge (phylum Porifera) growing over rocks. Note the irregular shape and numerous oscula.

Plate 10 Various worms.

(a) Ragworm (phylum Annelida, class Polychaeta, *Nereis* spp.) a free-living carnivorous worm. Note the segmented body and large parapodia.

(b)–(f) *Schistosoma* sp. (phylum Platyhelminthes, class Trematoda).

(b) Adult worms (10–20 mm long) in small blood vessels in the gut mesenteries of a laboratory hamster.

(c) The adult female worm is smaller and rounder than the male and lodges permanently in the groove along his body.

(d) A miracidium larva (150 μm long) that swims briefly in freshwater and parasitizes aquatic snails.

(e) The spined eggs of *S. mansoni* (150 μm long) lodge in small blood vessels in the wall of the bladder, from where they are excreted with the host's urine.

(f) A cercaria larva (1 mm long) that forms from other larval stages inside the snail. It may enter human tissues through a small wound.

▲ 11 (a)

▲ 11 (b)

▲ 11 (c)

▲ 11 (d)

11 (e) ▼

Plate 11 Phylum Arthropoda.

(a) Larva of a scarab beetle (class Insecta, order Coleoptera, family Scarabaeidae) showing the segmented body, spiracles, head, compound eyes, paired limbs and mouthparts. This diverse family of over 19 000 species includes some of the largest living insects.

(b) A tick (subphylum Arachnida, class Acarina) from West African rainforests. This specimen is engorged with a meal of vertebrate blood, making the abdomen huge relative to the limbs and eyes.

(c) A spider (subphylum Arachnida, class Araneae) from West African rainforests. Spiders have several pairs of simple eyes, pedipalps and chelicerae and eight legs. They are all predators; most eat other arthropods, particularly flying insects.

(d) A harvestman (subphylum Arachnida, class Opiliones). These fast-moving arachnids feed mainly on other arthropods.

(e) A tropical centipede (subphylum Uniramia, class Chilopoda). These fast-moving, carnivorous arthropods have a pair of long, agile legs on alternate segments of the body, long antennae and powerful mouthparts, sometimes including poisonous fangs.

(d) The smooth newt (class Amphibia, order Urodela, *Triturus vulgaris*). This female is swollen with eggs, which will be laid following a prolonged and elaborate courtship. The larvae are aquatic but the adults migrate long distances over land.

(e) A tree frog (class Amphibia, order Anura). The long hindlegs of adult frogs are specialized for leaping and climbing, although the tail is important for locomotion in the tadpole larva. Most adult frogs feed on arthropods, and are abundant and diverse in tropical and temperate forests. This brightly coloured species is probably poisonous.

(f) The Nile crocodile (class Reptilia, order Crocodilia, *Crocodilus niloticus*) sunning itself among papyrus reeds (subclass Angiospermidae, superorder Monocotyledonae, family Cyperaceae, *Cyperus papyrus*) in East Africa. All members of this ancient group are aquatic carnivores that lay eggs on land.

(g) Female (left) and male red-foot tortoise (class Reptilia, order Chelonia, *Geochelone carbonaria*) with newly hatched young. Like most chelonians, this South American rainforest species is omnivorous, eating plants, fungi, carrion and small animals.

(h) African chameleon (class Reptilia, order Squamata, family Chamaeleontidae, *Chamaeleo* sp.). Chameleons catch insects on the long, extensible tongue and are viviparous, giving birth to up to 20 babies at a time.

(i) American hognose snake (class Reptilia, order Squamata, family Colubridae, *Heterodon platyrhinos*). This species feeds on amphibians and large invertebrates and is not in fact venomous but, when threatened, it waves its tail and pretends to strike like the very poisonous rattlesnakes that occur in the same area.

Plate 16 Phylum Chordata, subphylum Vertebrata, class Mammalia.

(a) Foetal lions *in utero* showing the placentae and umbilical cords, through which all eutherian mammals obtain nourishment during gestation.

(b) All newborn mammals are nourished on milk that is produced only by the female. The duration of lactation is variable. Brown bears (order Carnivora, family Ursidae, *Ursus arctos*) are over a year old before they are weaned.

(c) Bats (order Chiroptera) are the second most diverse order of mammals and the only group that can fly. Most bats catch insects on the wing at night, and have acute hearing but reduced eyes.

(d) Hyraxes (order Hyracoidea, *Procavia* sp.) are a small order of herbivorous African mammals whose teeth and feet resemble those of elephants.

▲ 15 (d)

▲ 15 (e)

▲ 15 (f)

▲ 15 (g)

▲ 15 (h)

▲ 15 (i)

▲ 16 (a)

16 (c) ▼

▲ 16 (b)

16 (d) ▼

▲ 16 (e)

▲ 16 (f)

▲ 16 (g)

▲ 16 (h)

▲ 16 (i)

(e) Raccoons (order Carnivora, family Procyonidae, *Procyon lotor*) are common in North and Central America. They are nocturnal and feed on arthropods, carrion and plants both on the ground and in trees.

(f) Lions (order Carnivora, family Felidae, *Panthera leo*) are almost exclusively carnivorous. Unlike most felids, lions live and hunt in groups, often killing hoofed mammals several times their body mass.

(g) Seals (order Pinnipedia, family Phocidae) are powerful swimmers with short, stocky limbs and massive body muscles. Pinnipeds are most abundant and diverse in polar oceans. They always breed on land or on ice floes, although they feed in water, mainly on fish.

(h) American bison (order Artiodactyla, family Bovidae, *Bison bison*). Bovidae are a widespread and diverse family of hoofed, ruminant mammals that includes antelopes, wildebeest, sheep, goats and cattle. Most species live in herds, and adults of both sexes have horns that are not shed.

(i) The giraffe (order Artiodactyla, family Giraffidae, *Giraffa camelopardalis*) is a hoofed, ruminant mammal with simple horns that browses on tall trees in East Africa. Giraffes have acute hearing and smell.

(j) The African elephant (order Proboscidea, *Loxodonta africana*) is the largest living terrestrial mammal, eating on a wide variety of plant food both in forests and on grassland. The tusks are modified incisor teeth and the trunk is an extension of the nose.

(k) The howler monkey (order Primates, family Cebidae, *Alouatta palliata*) feeds on leaves, fruit and flowers in the rainforest of Central America. Like all apes and most monkeys, the eyes point forwards and the nose is reduced, and they live in troops. This species communicates by loud howling.

16 (j) ▼

▼ 16 (k)

Plate 17 Phylum Chordata, subphylum Vertebrata, class Aves.

(a) Pigmy owl (order Strigiformes, *Glaucidium* sp.) from Central America. Owls are widespread on all continents except Antarctica and are predators on arthropods and small vertebrates. Most owls are nocturnal, but this species hunts by day. The eyes are tube-shaped rather than spherical, so owls must turn the whole head to look sideways.

(b) King penguins (order Sphenisciformes, *Aptenodytes patagonica*) are native to the coast of Antarctica. Both the wings and feet are adapted for swimming and they catch fish under water. The small, fine feathers form a waterproof layer that is replaced regularly by moulting.

(c) African grey parrot (order Psittaciformes, *Psittacus erithacus*). Many birds, including almost all parrots, fowl, ducks and geese and some song birds, are herbivorous as adults, although they may feed their young on insects and other animal food. Most parrots are seed-eaters, and use the feet as well as the powerful beak to manipulate tough foods.

(d) African crowned cranes (order Gruiformes, *Balaerica pavonina*) live on grasslands and specialize in catching locusts and other orthopterans, which they startle by stamping the powerful feet. The elaborate plumage is fully developed only in breeding adults and is essential for courtship behaviour.

THE ORIGINS OF DIVERSITY ◆ CHAPTER 2 ◆

2.1 FUNCTIONAL AND COMPARATIVE ANATOMY

In Chapter 1, some of the great variety of structures and life cycles of organisms were described, and we explained how such features could be used as a basis for classification. This chapter is about techniques and concepts for investigating and explaining the natural role of the structures, habits and life cycles of organisms, both living and extinct. The implications of the biological relations of organisms to each other for their structure and physiology are also considered, and we introduce you to the kinds of evidence used to elucidate evolutionary relationships. Such information will help you to understand how and when the different groups of organisms arose, and some of the factors that determine the numbers and diversity of species and their range of habits and habitats.

An organism's anatomy can be interpreted from two points of view: functional anatomy, the interpretation of a structure in terms of the role that it plays in the organism's natural behaviour, development or physiology, and **comparative anatomy**, in which the structural resemblances between organisms are identified and described. Experience—and evolutionary and physiological theory—show that, with few exceptions, anatomically similar organisms also have similarities in biochemistry, physiology, behaviour, habits and life-style. Therefore, comparative anatomy is not only useful for establishing evolutionary relationships between organisms, but it provides the essential theoretical basis for the extrapolation of conclusions derived from experiments and observations of laboratory animals (such as rats and mice) to wild animals, domestic livestock and humans which, for practical and ethical reasons, cannot be studied directly.

Comparative anatomy is one of the longest established and most fundamental disciplines in biology. Much of the information used in taxonomic classification (see the Introduction to this book) comes from comparative anatomy. In theory, the properties of all organs and tissues should contribute to some extent to the assessment of the degree of similarity between organisms, but in practice shells, skeletons and other durable parts predominate in taxonomic classification because they are more easily examined and preserved. Structurally elaborate tissues such as ears, eyes, guts and brains have also been studied intensively by comparative anatomists.

◇ Which mammalian organs or tissues would you expect to be least thoroughly studied by comparative anatomists?

◆ Tissues such as adipose tissue (fat) that are variable in appearance and abundance and amorphous tissues such as blood (although the comparative biochemistry of blood has been extensively studied).

In spite of these limitations, comparative anatomy has been successful as a basis for both evolutionary and physiological studies, particularly in the case of organs for which the function is thoroughly understood. Thus the similarities in the structure of paws, teeth and guts of cats and dogs lead us to expect similarities in their diet, habitat, behaviour and life-style. However, there are also subtle but important differences that can be related to differences in habits.

◇ From your knowledge of cats and dogs, list some anatomical differences between them and relate the characters to differences in habits of wild members of the cat family (Felidae) and dog family (Canidae).

◆ Cats (and all other felids) have retractible claws, but dog claws are almost immobile. Cats extend their claws when climbing trees (and furniture!) and in grasping prey, but the claws are retracted during normal walking, so they wear much more slowly and remain sharp. Dog claws are always much more blunt so they are ineffective in climbing or in tackling prey. Cats have large, forward pointing eyes and ears, and find their prey mainly by sight and sound. Vision is less efficient in dogs, particularly at night, but they have long, elaborate noses and an excellent sense of smell.

The list could be extended, but you probably had to think quite hard to pinpoint and interpret the anatomical differences between these very familiar kinds of animal. As techniques for studying organisms improve, small, but functionally important, differences can be resolved more finely. Thus the muscles of cats and dogs are indistinguishable when examined with ordinary histological or chemical methods but more sophisticated techniques reveal differences in the biochemical mechanisms that supply energy to the muscles during exercise. Biochemical mechanisms capable of supporting prolonged exercise are much better developed in the muscles of dogs than in those of cats. This finding identifies the physiological basis for the familiar observation that dogs (and other canids) can gallop continuously for many minutes, but cats (and other felids) tire after less than a minute of maximal exercise. Wolves, foxhounds and wild hunting dogs can chase their prey to exhaustion, but lions, tigers and jaguars stalk the prey until they are near enough to reach it with a single dash or pounce. If the prey outruns them in the first few hundred metres, they abandon the chase. Techniques for relating structure to function are improving rapidly but, even so, some important characters still cannot be related to any particular anatomical structures or biochemical properties; for example, how would you deduce from dead specimens that waving the tail indicates aggression in cats, but appeasement or recognition in dogs?

Interpreting the anatomy of organisms from unfamiliar habitats, and that of less structurally elaborate organisms, such as bacteria and protistans, is much more difficult. For example, marine biologists have pondered the taxonomic status and possible habits of some worm-like, deep-sea animals called Pogonophora ('beard-carrying') since they were first discovered at the beginning of this century. Animals from the deep sea are very often dead by the time they reach the surface because the reduction in pressure badly distorts the tissues, so no one had ever seen a pogonophoran alive nor was anything known about their natural habitat. Figure 2.1 summarizes some of the observations that biologists made from examination of dead, probably damaged, pogonophorans. They are many centimetres long and very thin (in some species, the length is 300 times greater than the diameter) with from one to several hundred tentacles on one end, and groups of bristles at the other end. The coelom is divided into several compartments along its length but there are no true segments. The heart and blood vessels are relatively large, and there are plume-like tentacles that are extended by movements of coelomic fluid. There is no mouth and no gut, and most of the body cavity is occupied by an irregularly shaped sac called the trophosome.

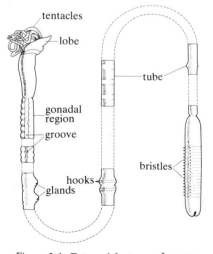

Figure 2.1 External features of a generalized pogonophoran.

◇ With this information, could you assign pogonophorans to any of the phyla described in Chapter 1?

◆ Not really. There are coelomate, worm-like animals in several phyla, including annelids, echinoderms (i.e. holothurians (sea-cucumbers)) and chordates, but the pogonophorans do not appear to have most of the other features that indicate that they should be classified in any of these groups.

The meagre evidence indicated that the most likely categories were Annelida or Chordata (see Chapter 1, Sections 1.6.3 and 1.6.7).

◇ What fundamental anatomical feature would distinguish an annelid from a chordate?

◆ In annelids, the nerve cord is ventral to the gut and the major blood vessels are dorsal, but the converse is true of chordates.

Unfortunately, even such seemingly simple facts could not be established for pogonophorans because, never having seen them alive, biologists did not know which side was dorsal or which end was anterior! Deep-sea pogonophorans were photographed alive in their natural habitat for the first time in 1977, when deep-sea submersibles and underwater cameras were used to explore volcanic activity on the ocean floor. The problem was resolved in a moment: the nerve cord is on the ventral side of the body and the tentacles are anterior.

◇ In the light of this information, how would you classify pogonophorans?

◆ The presence of a coelom, bristles and the position of the blood vessels and nervous system suggest that pogonophorans share the annelid body plan, although there are no true segments.

The biggest mystery was interpreting the functional anatomy. How could such a large animal sustain itself without a gut?

The use of deep-sea submersibles enabled scientists to take photographs and television pictures of the animals that live in deep-sea trenches near vents from which volcanic gases, such as hydrogen sulphide, emerge. Biologists were surprised by the abundance and variety of other animals living near the volcanic vents: there were bivalves, shrimps, crabs and other worm-like animals living in a totally dark environment in which plants, algae and other photosynthetic organisms were apparently absent. Among the largest and most spectacular was a pogonophoran named *Riftia pachyptila* that reaches a length of 2 m and a maximum diameter of 2 cm and consists of a pale body encased in a white papery tube, from which protrudes a large scarlet tentacle.

The clue to both the local abundance of animals and the pogonophoran's peculiar anatomy came when bacteria (see Chapter 1, Section 1.2.1) capable of metabolizing hydrogen sulphide, ammonia and other volcanic gases were found in large numbers near the volcanic vents. The trophosome was found to contain many such bacteria. Biologists now believe that adult pogonophorans are nourished by nutrients that these autotrophic micro-organisms synthesize from hydrogen sulphide and other inorganic compounds. The tentacles are scarlet because they contain large quantities of blood in which the concentration of a pigment similar to haemoglobin (see Chapter 1, Section 1.6.3) is unusually high. Pogonophoran blood pigment has the special property of combining with both oxygen and hydrogen sulphide at low concentrations. Gases dissolved in the water are absorbed through the walls of the tentacles, where they combine with the pigment in the blood and are conveyed by the well-developed heart and blood vessels to the bacteria in the trophosome. Several species of bivalves (see Chapter 1, Section 1.6.5), also found near volcanic vents, have greatly modified gills that harbour bacteria similar to those in pogonophorans and are probably nourished in a similar way.

2.1.1 Function and adaptation

The pogonophoran example shows that the study of the structure and properties of organisms has implications beyond providing a basis for their classification: it also helps biologists to understand their habits and the kinds of habitats that they would be expected to occupy successfully. The structure and physiological capabilities of organisms are said to be 'adapted' to their environment; particular organs or biochemical pathways suggest particular 'functions' within the organism. The concepts of **function** and **adaptation** are two of the most important concepts in both evolutionary biology and physiology so it is important to clarify exactly what we mean by these terms. Like any other concepts, they have to be used with caution, and can be misleading if used incorrectly. In this Section, we will illustrate some of the pitfalls in the interpretation of the natural function of biological structures and processes.

It is very tempting to use one's own experience to identify and explain adaptations. Thus we readily conclude that limbs are adapted for feeding and locomotion because these activities are important functions of our own arms and legs, but the role of limbs and extensions of the body wall acting as respiratory organs (Chapter 1, Section 1.6.4), is much less obvious to us mammals who take air into lungs through the nose or mouth.

◇ Suggest some ways in which the functions postulated for the nose and jaws might be different in a textbook of functional anatomy written by (a) elephants (Plate 16k) and (b) parrots (Plate 17c).

◆ Elephants might conclude that noses are essential for drinking and manipulating food, and more useful than forelimbs for carrying things and scratching themselves. Parrots might assume that jaws are essential for climbing and carrying things as well as for feeding and preening, and that fine manipulation of objects such as food and nest material is done by a beak and hind legs.

The observers' own experience influences their interpretation of the structure and adaptations in other organisms; only the wisest and most observant elephants and parrots would appreciate the importance to primates of the arms and hands in feeding, carrying things and climbing, particularly if they had never observed living specimens engaged in natural activities, such as feeding and nest building.

This example might seem trivial to you, but if you think about it carefully, you may realize that such **anthropomorphic** ('human form') thinking has led biologists both now and in the past into many erroneous interpretations of the form and function of animals and plants.

Another problem is that the relationship between form and function is neither necessarily unique nor exact. The skulls in Figure 2.2 all come from mammals classified in the order Carnivora*. Although, as the name implies, the

*Carnivora ('carnivores' in English) is the taxonomic name for an order of mammals, many of which are meat-eaters. 'Carnivorous' means any organism that feeds mainly or entirely on animals, regardless of its taxonomic affinities.

majority of species of Carnivora are predators on other animals (e.g. lions (Plates 16a and f), hyenas, weasels), some species have a mixed diet (e.g. badgers, raccoons) and a few (e.g. the giant panda) are herbivorous, feeding almost exclusively on plants. Figure 2.2a shows the teeth of a jaguar; the dagger-like canine teeth enable the animal to stab and tear its prey and the smaller incisors at the front of the mouth are suitable for cutting. Some of the molar and premolar teeth at the back of the jaw are large and elaborately shaped, while others are small, peg-like structures that do not meet (or occlude) when the jaws are closed. The former, called carnassials, act like scissors, slicing flesh from bones. The non-occluding teeth have no known function and are regarded as **vestigial** structures. They are variable in number and form, and are not present at all in some apparently normal specimens of the species.

The teeth of the tiger (Figure 2.2b), another member of the family Felidae ('cats'), clearly resemble those of the jaguar, and indeed the diets of the two species are broadly similar, although, being about twice the size, tigers take larger prey than jaguars. The timber wolf (Figure 2.2c) and the red fox (Figure 2.2d) belong to the family Canidae ('dogs') and are also primarily predators on other mammals and birds, although foxes sometimes eat invertebrates and a little fruit.

◇ List some similarities and differences between the teeth of felids and of the wolf.

◆ All specimens have long, stout canines, sharp incisors and pointed carnassial molars, but the canids have longer, narrower jaws and six molar teeth. The felids have only three molar teeth in each jaw.

The European badger (Figure 2.2e) is also a member of the order Carnivora but its molar and incisor teeth are flattened and are better suited to grinding than to slicing. It is omnivorous, eating a mixed diet of earthworms and other terrestrial invertebrates such as large insects and plant food, particularly grain. More than 90% of the diet of most bears (family Ursidae), such as the brown bear (Figure 2.2f and Plate 16b), consists of leaves, grass, fruit and seeds. Ursid canine teeth are long and sharp like those of other carnivores, but the incisor and molar teeth are broad and flat, and there is no indication of scissor-like action. Although the bear's teeth are much more worn than those of the canids and felids, tooth wear accounts for only a small part of the

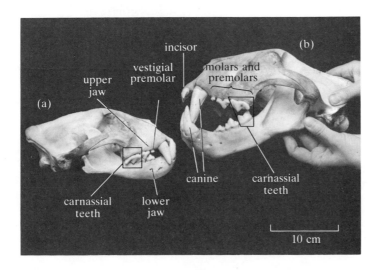

Figure 2.2 Skulls of various Carnivora: (a) Jaguar, *Panthera onca*. (b) Tiger, *Panthera tigris*. (Figure continued overleaf.)

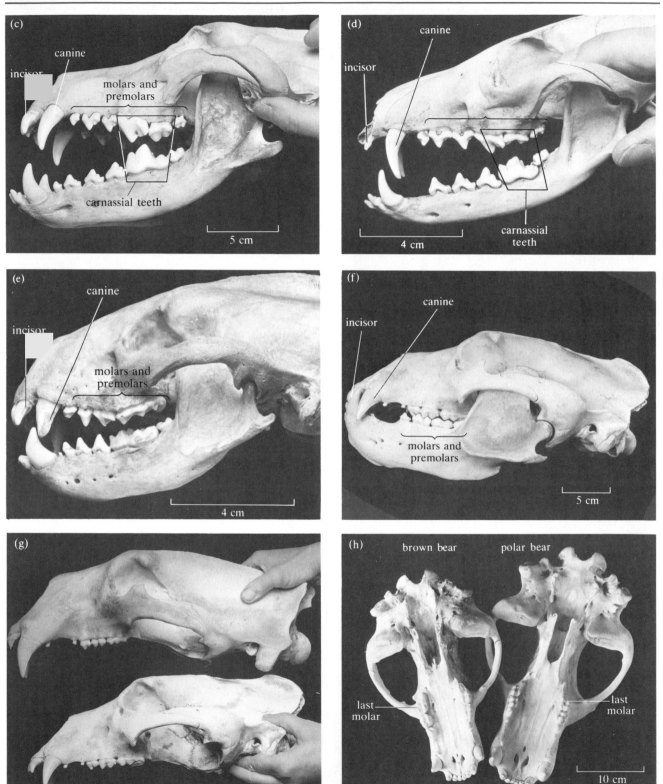

Figure 2.2 (c) Timber wolf, *Canis lupus*. (d) Red fox, *Vulpes vulpes*. (e) European badger, *Meles meles*. (f) European brown bear, *Ursus arctos*. (g) Cranium and upper jaw of a polar bear, *Ursus maritimus* (above) and the brown bear, *Ursus arctos* (below). (h) The upper jaw seen from below of the polar bear (right) and the brown bear (left). Note that the vestigial teeth are sometimes asymmetrical.

species differences in structure of the dentition. Herbivorous bears pluck grass and leaves using their front incisors rather like a horse, or harvest berries with their long, flexible tongue. They sometimes catch fish in shallow, fast-flowing streams by stunning them with their broad paws, and occasionally they eat carrion.

◇ From studying the form of its teeth, can you suggest (a) the taxonomic affinities and (b) the natural diet of the animal whose skull is shown at the top of Figure 2.2g?

◆ (a) The skull at the top of Figure 2.2g has a relatively long upper jaw, with three functional and fairly flat molar teeth and two very small vestigial molars in a large gap behind the canines. The similarities between this skull and that in Figure 2.2f suggest that it is an ursid. (b) As the comparison of the two skulls in Figure 2.2g shows, the teeth of the animal at the top are more like those of the brown bear shown at the bottom of the Figure than those of the felids or canids shown in Figures 2.2a–d, so it is probably an omnivore or herbivore.

Wrong! Figure 2.2g is a polar bear skull, and observations in the wild show clearly that these bears eat as much meat as tigers, jaguars or wolves. Figure 2.2h shows the upper jaws of the polar bear and brown bear from below.

◇ Compare the teeth of the carnivorous polar bear and herbivorous brown bear (Figures 2.2g and h) with those of the canids and felids (Figures 2.2a–d). Which features of the polar bear could be interpreted as adaptations to being carnivorous?

◆ The canines (Figure 2.2g) are relatively longer, stouter and more curved, and the molars, particularly the most posterior molar, are relatively smaller and more pointed in the polar bear than in the brown bear.

Nonetheless, the molars are not carnassial teeth and the polar bear dentition does not seem to be as fully adapted to a carnivorous diet as that of canids or felids.

What should we conclude from these facts? Polar bears are certainly not incompetent predators: they kill and eat seals (Plate 16g), many of which weigh more than 100 kg. However, unlike felids and canids, these carnivorous bears do not gnaw on bones in the side of the mouth. Instead, they nibble or tear on the food, using the incisor teeth on the front of the jaw. Carnassial teeth may be one of the many features that have enabled the cat and dog families to become widespread and (until hunted by humans) abundant predators, but they are not indispensible to being a successful carnivore. The study of fossil bears suggests that polar bears are a species that evolved relatively recently and became carnivorous during the Ice Ages. In fact, polar bears and brown bears have interbred successfully in zoos.

The skulls shown in Figure 2.3 illustrate another frequently encountered problem of interpreting the normal function of anatomical structures.

◇ From examination of the canine teeth alone, what would you conclude about the diets of the animals in Figure 2.3a?

◆ The left-hand skull in Figure 2.3a has massive canine teeth, so the animal is probably a predator; the canine teeth of the skull on the right are small and rounded, so the species is probably herbivorous.

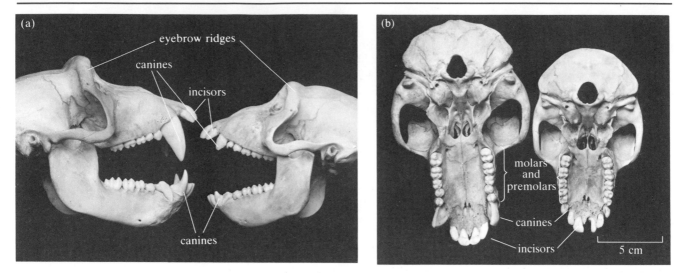

Figure 2.3 Skulls of pig-tailed macaque monkeys, *Macaca nemestrina*. (a) Side views. (b) The upper jaws of the same specimens seen from below. Left: adult male; right: adult female.

In fact, these skulls are from male (Figure 2.3a left) and female (Figure 2.3a right) specimens of the same species; they are macaque monkeys (order Primates) that feed mainly on flowers, fruit and leaves, although both sexes eat eggs, small animals and carrion from time to time.

◇What features of the rest of the dentition (Figure 2.3b, left and right) are consistent with a herbivorous diet?

◆The numerous molar teeth are broad and almost flat, more suitable for chewing than for slicing.

Study of the monkeys in the wild shows that the canine teeth function mainly as a threat display and, occasionally, in fights between males. The species is strongly **sexually dimorphic**, with fully grown males up to three times the size of adult females. Larger males with more massive canines are found to be more successful in securing dominance over troops of breeding females than smaller animals with less impressive teeth. The enlarged canine teeth seem to play little part in feeding, and may even be a hindrance in dealing with much of the normal diet.

◇What other features of the male's skull could be adaptations to appearing fierce?

◆The male's skull is much larger and has thicker, more prominent ridges over the eyes.

In this example, the information about other aspects of the monkey's structure, life cycle and habits discredits the initial conclusion, based mainly on the resemblance of its canine teeth to those of carnivores, that the left-hand specimen in Figure 2.3a is a predator. The study of just a few features of the anatomy and physiology of a species often leads to the wrong conclusions about their natural functions: to be interpreted correctly, individual structures and characters must be studied in the context of the biology of the animal as a whole. These examples also show how the study of anatomical or physiological features in the laboratory must be supplemented by observations of what the animals and plants actually do in the wild.

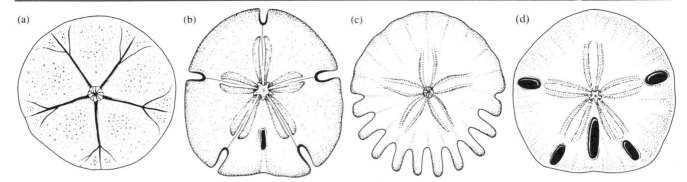

Figure 2.4 Adaptations of various sand-dollars to living in strong water currents: (a) *Echinarachnius*, with grooves on the ventral surface. (b) *Encope*, with notches evenly spaced around the edge of the body. (c) *Heliophora*, with deep indentations around the edge of the body. (d) *Leodia*, with lunules forming holes through the posterior half of the body. (a) is viewed from below. All the others are seen from above.

Most people would guess that the main function of teeth is in feeding, and even a role in intraspecific aggression is not totally unexpected, but the natural function of some structures seems to defy our imagination. Figure 2.4 shows several species of sand-dollars (see Chapter 1, Section 1.6.6) that are abundant in the Caribbean and along the east coast of America.

◇ What do you think are the functions of the grooves and notches on these sand-dollars?

Probably, you cannot produce any convincing suggestions. Neither, until recently, could biologists, most of whom had seen these animals only as dried or pickled specimens in a museum. Are the notches important for digging burrows, are they somehow involved in sexual reproduction, or are they simply a functionless accident of the way in which the animal develops? Careful observations of these species in the wild, combined with appropriate laboratory measurements, have demonstrated the role that these curious structures play in the animals' normal lives.

Sand-dollars live on, or buried in, sand, often in places that are exposed to waves and strong currents, where all but the densest objects may be swept away by the tide. Species that live in the most exposed sites tend to be flatter, and to have larger and more elaborate notches and lunules (Figure 2.4d). Some specimens have been examined in flow tanks similar to those used by naval engineers to investigate the forces generated by the movement of ships and submarines through water. As you might expect, flatter sand-dollars exposed to water currents remained in place longer than more domed species. But the biologists found that the notches and slits also played an essential role in protecting the animals from being swept away by water currents, probably by modifying the pattern of flow of water over the body. Normal adult *Leodia* remained in position in water currents of up to $0.4\,\mathrm{m\,s^{-1}}$ but if their lunules were artificially blocked, they were swept away by water movements of only $0.2\,\mathrm{m\,s^{-1}}$. Thus the body form of these sand-dollars is intricately adapted to the prevention of being swept away from suitable habitats by strong currents.

It is almost impossible to prove that a feature is completely devoid of any 'function'. A great many features that were at first thought to be vestigial structures, developmental anomalies or 'ornamentation' have, on more thorough investigation, turned out to be adaptations to aspects of the organism's life with which we, as humans, are not familiar and therefore do not take into consideration. Unfortunately, the criteria for classification (see the Introduction to this book) were, in many cases, established before the organisms' biology was thoroughly understood, so some of the key taxonomic characters are not necessarily important functionally. Conversely, functionally important features may be of only minor significance to taxonomists. The moral of the story is that biologists' understanding of the 'function(s) (if any)' of a structure or biochemical process depends upon a comprehensive knowledge of the organism's natural habits and life cycle. Suggestions about function remain theories until they can be confirmed, where possible, by detailed observation of the organism in the wild, or tested experimentally.

Teleology

'Birds have wings in order to fly'. 'The annelids evolved a coelom in the mesoderm to burrow into compacted sand or soil more effectively than platyhelminths could do.' 'Humans have subcutaneous fat to insulate them against the cold'. All these statements refer to facts; birds do have wings and they do fly; annelids do burrow into denser substrates than platyhelminths do, and some people have extensive superficial fat and can endure prolonged exposure to cold. Nonetheless, biologists would find fundamental flaws in the way in which the anatomical and physiological facts are related to their functions in these statements because they go beyond the facts and assert a false relationship between cause and effect. Statements implying that structures or processes evolved in order to achieve a certain goal are called teleological. **Teleology** is outlawed by biologists, not only because it implies an erroneous theory about the mechanism of evolution, but also because, as we saw earlier in this section, it may cause people to jump to hasty, and often incorrect, conclusions about the actual roles that structures and tissues play in the organism's natural activities.

The first statement implies that the ancestors of birds, prompted perhaps by watching insects flying around, thought 'flying looks like a comfortable way to get around, so let's acquire wings and fly'. But anatomical and physiological changes normally evolve gradually by the accumulation of modifications of structure that, when they first appear, are usually unrelated to the function that they eventually assume, or indeed to any particular function. Annelids, frustrated by the difficulties of penetrating compacted soils, did not 'decide' to extent their range by evolving a body plan that is better suited to burrowing. There is no evidence that the course of evolution is directed by need or ambition.

Nonetheless, many people, including, unfortunately, some biologists, use teleological thinking as a shorthand way of describing adaptations. For example, the features of the dental anatomy and feeding habits of the carnivores described earlier in this section could be summarized thus:

> Although handicapped by being descended from omnivorous ancestors, polar bears are trying to evolve sharp cusps on their molar teeth in order to slice meat from carcasses in the same way as tigers, jaguars and wolves do. In the meantime, they make do by using incisor teeth to tear their food.

Biologists would present the same information as follows:

> The cat family are all carnivorous predators and have short, powerful jaws in which the molar teeth are sharply pointed and slice meat with a scissor-like action. Most bears feed mainly on plants and have longer jaws with several broad, flat molar teeth. Polar bears are exceptional in being almost entirely carnivorous and their functional molar teeth are more pointed than those of other bears. In these respects, the molar teeth of polar bears resemble those of felids and may play a similar role in shredding meat. However, most of their normal feeding movements are similar to those of the other bears.

The second version is preferable because its implications about how the situation might have evolved are correct. Although modern humans may design and build clothes, vehicles and other apparatus in order to colonize unfamiliar environments, the ancestors of polar bears did not sit on some subarctic shore and say to themselves, 'plants are becoming scarcer, and there are lots of delicious seals up there, so let's copy the tigers' and jaguars' example and evolve pointed molar teeth for eating meat'. Biologists would explain the evolution of carnivorous polar bears from omnivorous forest-living ancestors thus:

> Brown bears were widespread over the whole of Europe, northern and central Asia and North America. The proportion of plant food, fish and carrion in the diet varied with the season and geographical range of the population. There was less plant food available to bears living in colder climates in which there was extensive ice and snow cover. Bears with more pointed molar teeth were more efficient at catching fish and shredding meat, although less efficient at grinding tough leaves and grass. As food availability changed, bears that were better suited to meat-eating were more likely to prosper and breed successfully than those whose teeth were adapted to chewing plants. Genetic or developmental mechanisms that produced sharper cusps on the molar teeth were favoured by natural selection and bears with these features became more abundant than those lacking them.

According to this scenario, the evolutionary origin and present function of sharp molar teeth in polar bears had little direct connection with the occurrence of similar carnassial molars in the cat family.

Summary of Section 2.1

Nearly all biological structures are adapted to some function, but detailed study of the entire organism in its natural situation is nearly always necessary to identify the natural function correctly. Statements about biological structures or processes that imply purpose or directed creation are not appropriate to scientific thinking and writing.

Question 1 (*Objective 2.2*) Outline some methods of identifying the natural function of a structure.

Question 2 (*Objective 2.3*) Which of the following statements are teleological? Rephrase the teleological examples in a way that expresses the same meaning but avoids teleology.

(a) Polar bears have become carnivorous because there isn't much plant food in the Arctic in wintertime.

(b) In many species, males must be larger and have stronger teeth to protect their females from predators and rival males.

(c) Cats retract their claws to keep them sharp.

(d) Some sea-urchins have poisonous spines that protect them from predators such as fishes.

(e) Tall trees grow long, spreading roots to avoid being blown down by strong winds.

(f) To avoid desiccation, animals such as newts venture onto land only at night.

(g) Sea-anemones have a large mouth to swallow large prey.

(h) Mosses use the rain to disperse their gametes.

2.2 THE EVOLUTION OF DIVERSITY

With very few exceptions, it is impossible to establish the ancestry and exact course of evolution of a single species, much less of a larger taxonomic group. All theories about the origin of particular characters and interrelationships between species are based upon inferences from comparison of the structure, development and habits of organisms, both living and extinct. One important source of evidence, comparative anatomy, has already been discussed (Section 2.1). This Section is about some other kinds of evidence that contribute to our understanding of the origin of biological diversity.

2.2.1 Life cycles

As mammals, we tend to think of juveniles as miniature adults, similar in body form and habits to their parents. However, it is important to realize that mammals, birds and reptiles are exceptional in having **direct development**. In most animals, including the majority of species in such abundant groups as bony fishes, molluscs, echinoderms, platyhelminths, annelids and arthropods (see Chapter 1, Section 1.6), the juveniles are sufficiently different in structure and habits from their parents to be regarded as **larvae**. Larvae normally live independently of their parents and feed for themselves, and, with a very few exceptions, they never reproduce sexually; indeed, often, as with the eft larva of newts and with most caterpillars, it is impossible to establish which sex they will develop into. In many cases (e.g. barnacles, many molluscs and echinoderms), the larval stage is brief compared to the longevity of the adult stage, but some species spend most of their lives as larvae. For example, the insect order Ephemeroptera ('day wings'; mayflies) was so named because the adult stage lives for only one day, although the larvae may have spent several years in freshwater, where they feed on detritus and small invertebrates.

◇ What kinds of animals would you expect to have larvae?

◆ Species that are sedentary as adults (e.g. parasites); those that exploit seasonally or locally abundant foods (e.g. tadpoles, fish larvae).

A huge variety of planktonic larvae exploit the transient abundance of bacteria, algae—and each other. The abundance of both marine and freshwater plankton often fluctuates due to seasonal changes in sunlight and in the availability of inorganic nutrients. Some larvae (Figure 2.5) are similar in general appearance to the adults that they will become, but others such as the veliger larva of marine gastropods (Figure 2.6) are completely different.

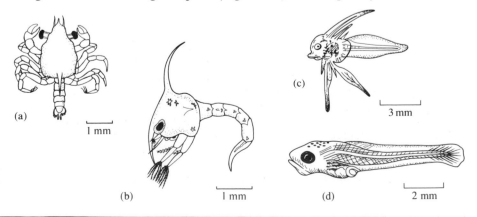

(a)

1 mm

(c)

3 mm

(b)

1 mm

(d)

2 mm

Figure 2.5 Some marine planktonic larvae: (a) and (b) Crustacean larvae. (c) and (d) Fish larvae.

◇List some resemblances and differences between the veliger larva (Figure 2.6) and adult marine gastropods (see Chapter 1, Section 1.6.5, Figure 1.26).

◆Both larval and adult gastropods are roughly bilaterally symmetrical and have a shell, but the larval foot is ciliated, whereas the adult foot is muscular.

The larvae of most marine molluscs are planktonic, feeding on algae, micro-organisms and other eggs and larvae. When they reach a certain size, they settle to the bottom and develop adult characteristics. Almost all larval echinoderms are also planktonic (Figure 2.7). Unlike the adults, the larvae do not show pentamerous symmetry and they do not have tube-feet. They move by means of bands of cilia over the body and its projections. After spending some time in the plankton, a rudimentary skeleton begins to form and the larvae sink to the sea-bed where they develop the adult body shape and internal structure.

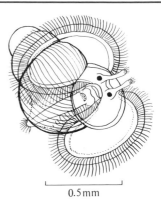

Figure 2.6 Veliger larva of a gastropod. Note the paired eye-spots and tentacles, ciliated foot and rudimentary shell.

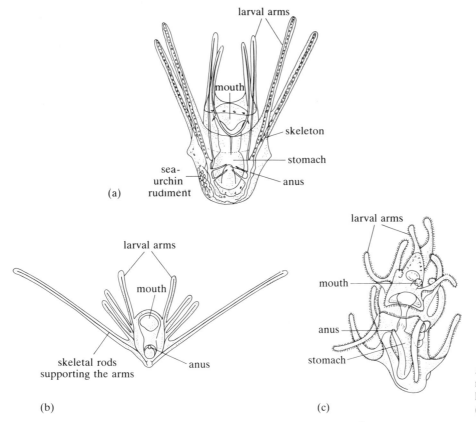

Figure 2.7 Echinoderm larvae: (a) Echinoid. (b) Ophiuroid. (c) Asteroid. All larvae are approx. 0.5 mm in diameter (excluding arms).

Insects have two main kinds of life cycles. The larvae of exopterygote insects (Figure 2.8a) have visible wing pads, although only the adults have functional wings and the period of inactivity around the time of the final moult is not much longer than that between the larval moults. The orders Ephemeroptera, Odonata (dragonflies and damselflies), Orthoptera ('straight wings'; grass-hoppers and crickets) and Hemiptera ('half wings'; bugs, aphids and cicadas) and several less familiar orders such as Anoplura (lice; see Section 2.4.3) are exopterygotes. Orthopteran larvae are similar to the adults in general body form and often in diet and habits but the aquatic larvae of mayflies, dragonflies and damselflies are quite different from the adults in both appearance and habits.

More than 80% of living species of insects, including Coleoptera ('sheath wings', beetles), Lepidoptera ('scaly wings', butterflies and moths), Hymenoptera ('united wings', wasps, bees and ants), Diptera (flies), and a few less familiar orders (e.g. caddis-flies, fleas) are endopterygotes. Endopterygote larvae (Figure 2.8b) include caterpillars (lepidopterans), grubs (coleopterans and several other groups) and maggots (dipterans); they never have externally visible wing pads and are nearly always very different from the adults in body form, and usually also in diet and habits. There is a resting stage, called a pupa or chrysalis in lepidopterans, that lasts from a few hours to many months before the adult emerges from the final moult.

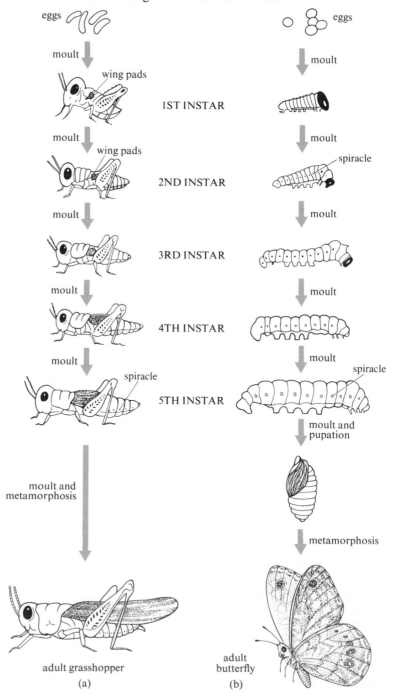

Figure 2.8 Exopterygote and endopterygote development in insects. (a) Grasshoppers (order Orthoptera) have exopterygote larvae in which the developing wings are visible and enlarge with successive moults, and no pupal stage. (b) Butterflies (order Lepidoptera) have endopterygote larvae which are very different in form and habits from the adult, and metamorphose into the adult in a pupa.

◇ Name an important contrast between the role of insect larvae and those of echinoderms and molluscs in the life cycle of the species.

◆ Most adult echinoderms and molluscs (e.g. gastropods, bivalves) do not move very far or very fast but, because the larvae may drift long distances in the plankton, the species is dispersed mainly at the larval stage. Most adult insects can fly, but the wingless (sometimes also legless, e.g. maggots) larvae are relatively immobile, so the adult is the main dispersal stage.

The great majority of caterpillars and many beetle grubs feed on angiosperm or gymnosperm plants, particularly newly formed buds, leaves and roots. A particular species of insect can usually only flourish on one or a few species of plants, and under favourable conditions a heavy infestation of larvae can defoliate large trees in a few days. Many dipteran larvae (maggots) feed on carrion although a few are parasites (see Section 2.4.2), living in or on the skin of other animals. Some hymenopterans are social, living in large colonies in which the older individuals feed the larvae. The larvae of many species of solitary wasps are parasitic: the mother lays one or several fertilized eggs in or on another insect larva (usually a caterpillar or a grub). The eggs hatch and the parasitic larvae gradually eat the host's tissues, but may not actually kill it until the parasites are large enough to pupate and emerge as free-living adult wasps. The life cycles of many dipteran flies are similar to those of these solitary wasps.

Larval forms in evolution

As well as contributing to an understanding of the biology of the species as a whole, the structure and habits of larvae can also provide clues about an organism's ancestry and evolutionary history. Sometimes (but by no means invariably), the larvae of species descended from a common ancestor resemble each other (and/or the ancestral form) more closely than do the adults. Thus invertebrate larvae, including those shown in Figures 2.6, 2.7 and 2.8b bear little resemblance to the adult forms. Larvae of the same taxonomic group often have features in common that distinguish them from the larvae of other groups. But similarities between adults do not always correspond exactly to resemblances between the larval stages. Thus adult starfish (see Chapter 1, Section 1.6.6, Figure 1.29), resemble adult brittle-stars (see Figure 2.12) more closely than they do sea-urchins and sand-dollars (see Figure 2.4), but echinoid (Figure 2.7a) and ophiuroid larvae (Figure 2.7b) are similar and differ from larval asteroids (Figure 2.7c).

The study of larval forms has been particularly important for unravelling the major events in the evolutionary history of our own phylum, the Chordata (see Chapter 1, Section 1.6.7). Sea-squirts (Figure 2.9a) are urochordates that are common on pilings and rocks below the shoreline; some species grow to a height of several centimetres.

◇ Which features of the *adult* sea-squirt (Figure 2.9a) are (a) consistent (b) inconsistent with its classification as a chordate?

◆ (a) It has a large pharynx perforated by numerous gill slits. (b) It lacks a notochord, the nervous system is diffuse and the arrangement of the gut and heart are not typical of chordates.

In other words, the adult anatomy of sea-squirts provides little compelling evidence that they should be classified as chordates. However, the sessile adults develop from a free-living larva (Figure 2.9b).

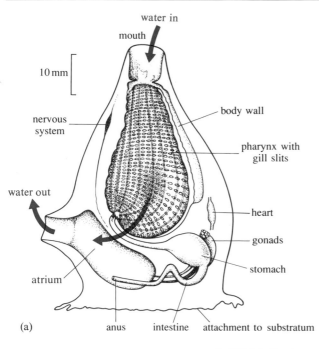

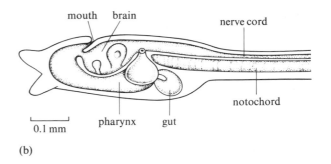

Figure 2.9 Urochordates: (a) Adult sea-squirt (cut in half and seen from the left side). (b) The major features of a larval urochordate, the ascidian 'tadpole'.

◇ Which features of the *larval* sea-squirt (Figure 2.9b) are (a) consistent (b) inconsistent with its classification as a chordate?

◆ (a) The larva has a notochord, a tail and a clearly defined dorsal nerve cord. (b) The position of the mouth is atypical.

The tiny ascidian larva has more chordate features than its parent. It lives in the plankton and swims actively with a movement resembling that of an amphibian tadpole. The two other groups of urochordates are more abundant, although less familiar because they live in the ocean plankton. They have contrasting life cycles: the barrel-like salps resemble adult sea-squirts (Figure 2.9a). They have no larva and develop directly into miniature adults. The appropriately named class Larvacea (Figure A.40) also secrete a gelatinous covering around themselves and float freely in the plankton but at all stages of their life cycle, their basic anatomy is similar to that of an ascidian larva. The salps may have evolved by the reduction and eventual elimination of the larval stage, while the Larvacea seem to be 'larvae' that have acquired the ability to produce gametes and reproduce sexually. The appearance of species in which the adults resemble the juvenile stages of other species is called **neoteny**, and may be a widespread process in the evolution of new groups.

◇ Do typical adult cephalochordates such as amphioxus (Figure 2.10) and vertebrates (Chapter 1, Figure 1.31) resemble adult or larval sea-squirts (Figure 2.9)?

◆ Adult cephalochordates and vertebrates resemble larval sea-squirts in having elongated body shape, a prominent internal skeletal rod, tail and an active swimming habit.

In almost all living species of vertebrates, a notochord-like structure appears in the early embryo and is gradually replaced with hard skeletal materials such as bone. By the time the animal becomes free-living, the vertebral column is the principal component of the skeleton. Nonetheless, these

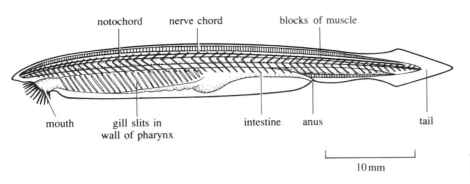

Figure 2.10 Amphioxus, a cephalochordate (cut in half and seen from the left side).

similarities in developmental anatomy and body plans suggest that vertebrates may have originated from larval urochordates, or from some common ancestor of both urochordates and vertebrates.

◇ Using this principle, can you suggest what aspects of the embryonic development of pogonophorans (Section 2.1) could be investigated to establish the origin of the trophosome?

◆ The inner lining of the gut is always endoderm (see Chapter 1, Figures 1.18, 1.19a, 1.20 and 1.21). If the trophosome evolved from organs lining the coelom, it would develop as part of the mesoderm, if it is a modified gut, part of it should develop from the endoderm.

Studies of the development of the trophosome show that it does indeed develop from endoderm, supporting the conclusion that the trophosome is a greatly modified gut.

The life cycles of most non-flowering plants also involve several independent and distinct stages, although the immature stages are not larvae as defined for animals. Nonetheless, the life cycles of land plants are regarded as so fundamental to their evolutionary origins that this feature is a major criterion for their classification (see Chapter 1, Section 1.4).

2.2.2 The fossil record

A fossil can be the petrified remains of an organism's body, or a trace or relic produced by an organism, such as a burrow, footprint, moulted cuticle or excreta. The conditions for fossilization in most sediments are such that only hard parts such as shells and skeletons are preserved. Consequently, there are very few fossils of soft-bodied organisms such as worms and fungi, and of most kinds of parasites. The fossil record of most kinds of micro-organisms is also scanty, although some shelled sarcodine protistans, such as foraminiferans and radiolarians, are very abundant as fossils and have been extensively studied because they are associated with the presence of oil. However, there are a few kinds of rock, notably those containing a high proportion of phosphates, in which fossils are so well preserved that the detailed structure of internal organs, and even of cells and intracellular organelles, is clearly visible. Studies of the form and geological origin of fossils provides information about both anatomical structure and where, when and in association with which other fauna and flora an organism was living. Fossils are found in sedimentary rocks, particularly those formed in lakes, rivers and shallow seas. Consequently, there is far more information about ancient organisms that lived in such habitats than about oceanic, mountain or desert species which were rarely preserved.

Fossils are not only the sole source of information about the structures of extinct organisms and the parts of the world in which they lived, they also provide information about the minimum length of time for which major groups have been distinguishable and about the course of evolution. Evolutionary theory is based upon historical information from fossils and observations on the structure and genetics of living organisms. As you might expect, we know much more about living organisms than about extinct species but fossils are the major source of evidence for theories about the course of evolution and the origins and interrelationships of the major groups of animals and plants described in Chapter 1.

The fossil record for most lineages of organisms is too fragmentary to provide many details about the course and time-scale of evolutionary change.

◇ In which group would you classify the fossil in Figure 2.11?

◆ The large compound eyes clearly indicate that it is an arthropod but the general shape of the body and head do not suggest close affinities with any living class.

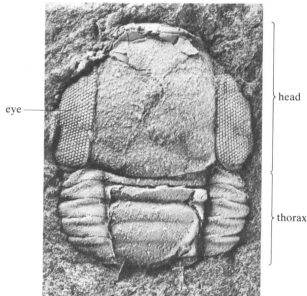

eye — head

thorax

Figure 2.11 An Ordovician trilobite from shales near Builth Wells in central Wales. The head and thorax are clearly visible, and the detail of the eyes is remarkably well preserved, but the limbs and the posterior part of the body are missing. Most trilobites are only a few centimetres long, but a few species became much larger. The head of this specimen is 1 cm across.

Figure 2.11 is a trilobite, one of the most abundant and diverse kind of fossils that are common in aquatic sediments throughout the Palaeozoic. They are among the earliest arthropods to appear in the fossil record, and there is no indication that they were ancestral to any major living group, so their relationships to other arthropods are uncertain.

◇ Referring to Chapter 1, Section 1.6.4, can you suggest why the remains of arthropods should be particularly abundant as fossils?

◆ Arthropods grow by moulting and the moulted cuticle of the juvenile stages is often durable enough to be fossilized, particularly if it is calcified, and is of sufficiently complex structure to be readily identifiable. Many arthropod fossils are the remains of moulted cuticle as well as the whole body.

Details of joints, spiracles and sense organs such as eyes and antennae are sometimes almost as visible in fossilized arthropod cuticle as in the exoskeletons of living species. Shallow coastal seas and estuaries are favourable locations for fossilization, and harbour abundant and diverse arthropod faunas so the fossil record of aquatic arthropods is particularly rich. In certain shales in the Welsh mountains, both fossilized whole corpses and shed cuticles are so numerous and the preservation is so good that palaeontologists have been able to document gradual, progressive changes in the structure of the eye and dorsal surface of the body over millions of years, as predicted by Darwin's theory of evolution. Nonetheless, other parts of the body, notably the legs and mouthparts, were very rarely preserved, so palaeontologists are still unsure about some elementary aspects of the biology of these animals, such as how they fed and moved.

◇ Suggest an explanation for why the limbs are absent from fossils in which other structures are very well preserved.

◆ The limbs could have been composed of very soft cuticle that decays rapidly after death. Predators could have eaten the limbs after death. Or perhaps many species were limbless.

Trilobites disappeared from the fossil record in the early Mesozoic, so unless more fully preserved specimens are found, we shall probably never know exactly how they lived.

Many groups of organisms, e.g. trilobites and some groups of reptiles (see Chapter 1, Section 1.6.7), that were very abundant and diverse in the past are now completely extinct, or reduced to few species. Did they die out because the climate became unfavourable or because their food supply disappeared, or was their disappearance the result of some major ecological catastrophe? Often we can only guess at the factors that brought about their decline but, in some cases, integrating knowledge of living species with the fossil record of their extinct relatives enables us to be more precise.

Echinoderms (Chapter 1, Section 1.6.6) have an extensive hard skeleton and their fossil record is one of the longest and most complete. Several classes are completely extinct, and some others, notably crinoids (feather stars) and ophiuroids (brittle-stars) were very abundant as fossils particularly in the Palaeozoic, but are minor components of the modern fauna. Brittle-stars owe their name to their habit of breaking off one or more of their arms when mauled by a predator. The discarded limb continues to wriggle for several

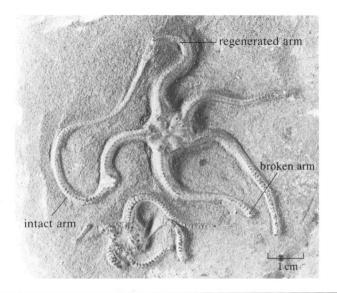

Figure 2.12 Ophioderma egertoni from rocks of Lower Jurassic age near Lyme Regis in Dorset. In undamaged ophiuroids, all five arms are long and tapering, like the one on the left of the large specimen. The shorter arm at the top was probably broken and regenerated during the animal's life. There is no sign of regeneration of the other arms, so breakage probably occurred at or after death.

minutes, and may distract the predator for long enough for the rest of the animal to escape. The stump regrows, but the skeleton of the regenerated arm differs from that of the intact limb. In many fossil ophiuroids, such as those in Figure 2.12, palaeontologists can distinguish regenerated arms from intact arms, enabling them to assess the level of predation on these ancient populations.

◇ Why is counting regenerating limbs a better measure of predation than simply counting the number of broken arms?

◆ Because breakage of the skeleton could have occurred after death. If the arm is even partially regenerated, the animal must have been alive and survived the injury.

In deposits of Jurassic age and earlier, some of which contain the remains of thousands of brittle-stars, only 2% or less of the specimens show signs of regenerating arms. But from the early Cretaceous, ophiuroids become much less common as fossils, and a higher proportion of the specimens have regenerating arms. Brittle-stars are now quite rare except on the floor of deep oceans, where there are few large predators, and in most modern populations 70% or more of the adult specimens have lost and regenerated at least one arm.

◇ What can you conclude about the decline of brittle-stars since the Cretaceous?

◆ Increased predation may have led to the extinction of many ancient populations.

Figure 2.13 Summary of the fossil record for terrestrial plants (Embryophytina). The widths of the pink areas are proportional to the numbers of species found in rocks of that age.

Evolutionary change

Many, but by no means all, fossils are sufficiently similar to living species to be assigned to modern groups of organisms. Figure 2.13 and Table 2.1 are summaries of the occurrence and relative abundance of some of the groups

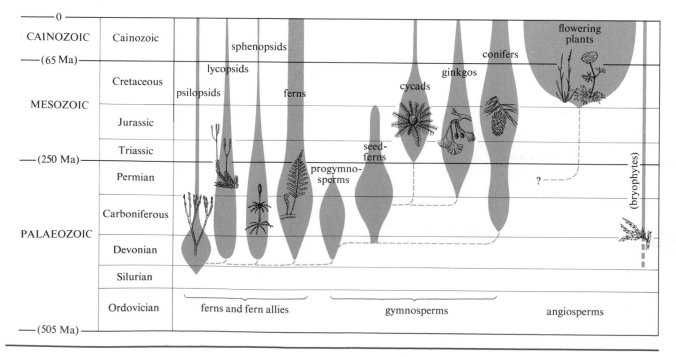

and organisms described in Chapter 1, based upon the fragmentary informa-tion from the fossil record. As you might expect, prokaryotes were abundant long before eukaryotes appeared, and unicellular algae were numerous and diverse before there were any multicellular plants. Although many of the major groups appear during the Devonian, other important kinds of plants, notably the flowering plants, first become abundant as late as the Cretaceous.

Table 2.1 The fossil record (in evolutionary sequence, from bottom to top, not drawn to scale in time).

Era	Period	Millions of years (Ma) ago	Plants	Animals
Cainozoic	Quaternary		major plant migrations and extinctions in response to Ice Age climatic changes	modern species of mammals and birds; later dominated by human societies
		2		
	Tertiary		angiosperms abundant in extensive forests; later many herbaceous species appear	placental mammals and birds more abundant; third diversification of insects
		65		
Mesozoic	Cretaceous		radiation of angiosperms, especially woody species: gymnosperms dwindle	most large reptiles extinct by end of period; diversification of teleosts
		145		
	Jurassic		gymnosperms dominant (conifers, cycads, ginkgos); angiosperms appear at end of period	reptiles abundant; first birds and archaic mammals; first teleosts late in period; more insects
		215		
	Triassic		many ferns and gymnosperms (seed-ferns and conifers)	large reptiles: first dinosaurs, turtles, ichthyosaurs, plesiosaurs appear; first mammals late in period
		250		
Palaeozoic	Permian		sphenopsids and ferns dominant; gymnosperms increase; advanced bryophytes	reptiles more abundant and replace amphibians on land; teleost ancestors diversify in sea
		285		
	Carboniferous		lycopsids, sphenopsids and ferns dominant; gymnosperms (seed-ferns) present; swamps lead to coal deposits	amphibians abundant; first reptiles on land; bony fishes in swamp pools; sharks in sea; first diversification of insects
		360		
	Devonian		early land plants; many psilopsids; primitive bryophytes	fishes abundant (mostly in freshwaters); first amphibians
		410		
	Silurian		first land plants (*Rhynia*)	first insects (on land); primitive vertebrates (mainly in freshwaters)
		440		
	Ordovician		marine algae	first vertebrates; molluscs (cephalopods) and brachiopods dominant in sea
		505		
	Cambrian		marine algae	most of the major groups of invertebrates, in sea
		590		
Proterozoic	Precambrian		multicellular plants and animals present 700 Ma ago eukaryotes present 1 000 Ma ago prokaryotes present 3 400 Ma ago	
		4 600		

The fossil record of animals presents a more surprising picture. Most of the major phyla of animals appear in the fossil record at about the same time, at the base of the Cambrian, and thereafter few completely new body plans develop. There are a few genera of organisms, notably *Lingula*, a shelled brachiopod that superficially resembles a mussel, that have persisted almost unchanged since they first appeared in the fossil record in the Cambrian. Table 2.1 contains information about only those groups of animals that include significant numbers of living species. There were many more phyla that appeared early in the Cambrian and are now almost or entirely extinct, but we do not have space to discuss them. Fossils are the only means we have of knowing whether a particular organism existed at a particular time and place. In the absence of information to the contrary, we have to conclude from Figure 2.13 and Table 2.1 that all the earliest organisms were marine, and that for a long time, almost half the total time for which life has existed on Earth, all organisms lived in water. However, it is impossible to determine whether all these contrasting groups of animals did indeed appear in the sea more or less simultaneously at the beginning of the Cambrian, as the known fossil record suggests, or whether they evolved over a much longer period and in other environments, but failed to form fossils.

Vertebrates are almost unique among the major animal groups in that nearly all their earliest representatives are found in rocks formed in ponds and rivers, indicating that much of the early evolution of the major kinds of fishes took place in freshwater.

◇ What difficulties do these facts present for the theory of the origin of vertebrates described in Section 2.2.1?

◆ Urochordates and cephalochordates are, and so far as we can tell from the fossil record always have been, marine (see Chapter 1, Section 1.6.7).

By bringing together information about the numbers and diversity of species of all the different kinds of organisms, and integrating it with interpretations of their habits, we can reconstruct possible ecological relationships between fossil species.

◇ From Table 2.1, which kind of animal is likely to have brought about the decline of brittle-stars since the Cretaceous, described earlier in Section 2.2.2?

◆ Teleost fishes would be high on the list of suspects, although it is not yet possible to establish for certain that they were the predators.

2.2.3 Relationships between phyla

The pink dotted lines and question marks on Figure 2.13 indicate possible interrelationships between the major groups of plants. The question marks are there for a good reason: direct evidence for evolutionary lineages is nearly always weak, and comes mainly from the study of similarities between living forms. There are very few fossils, some palaeontologists would say none, that have characters that indisputably qualify them as intermediates between two animal phyla or divisions of the plant kingdom.

One of the most famous 'links' between two phyla is a genus of elongated animals called *Peripatus* (Figure 2.14 and Plate 12) that are found in leaf litter and rotting wood in sub-tropical rainforests in several parts of the world, including Australia, southern Africa and Central and South America. *Peripatus* has features of the body plans of both arthropods and annelids. Like annelids, *Peripatus* has a central gut and layers of circular and longitudinal

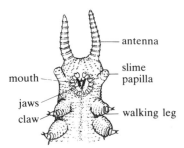

Figure 2.14 The anterior end of *Peripatus*, seen from below.

muscles. The body wall is soft and deformable and it can change its body shape and squeeze through very narrow crevices. However, the outer covering of *Peripatus* is cuticle, similar in structure and chemical composition to that of arthropods, and it has a haemocoel and simple tracheae. The cuticle is soft and flexible over most of the body, but it has stout claws on the limbs, and hard mouthparts. The thin cuticle of both the adults and the eggs is also fairly permeable to water, and *Peripatus* is confined to damp habitats. *Peripatus* falls outside the definition of both the Annelida and the Arthropoda and so most authorities classify it and a few similar genera in a separate phylum, the Onychophora (claw-carrying).

To have been a link between the annelids and arthropods, onychophorans would have to have lived early in the Cambrian, about 600 Ma ago. Unfortunately, the fossil record gives us little information about when and where onychophorans lived. There are very few fossil onychophorans, none of them old enough to prove that the group existed before the arthropods appeared in the fossil record. *Peripatus* can only tell us what a transitional stage in the evolution of arthropods from annelids might have been like. Its existence tells us very little about whether or when such creatures were actual descendants of early annelids and ancestors of the first arthropods.

2.2.4 Adaptive radiation

On current estimates, there are about 45 living species of pogonophorans (Section 2.1), 80 of ctenophores (Chapter 1, Section 1.6.2), 40 of Sphenopsida, all of them in the single genus *Equisetum* (horsetails) (Chapter 1, Section 1.4.2) but over 23 000 of bryophytes (Chapter 1, Section 1.4.1), 80 000 of nematodes (Chapter 1, Section 1.6.3) and over 100 000 different species of molluscs (Chapter 1, Section 1.6.5). By far the most abundant group of organisms is the arthropods (Chapter 1, Section 1.6.4), of which there are more than a million species, 400 000 of them beetles (Order Coleoptera).

Most people recognize the major differences between snails, slugs, oysters, mussels, octopus and squids but it takes an expert to identify the major categories of roundworms and algae. In other words, the species in these groups, although very numerous, are similar in structure and habits. In phyla such as the annelids, arthropods, molluscs and echinoderms, variations and elaborations of the basic body plan have given rise to creeping, burrowing, swimming, planktonic and encrusting forms. This process of diversification into several very different lineages is called **adaptive radiation**. There are freshwater and terrestrial annelids, arthropods and molluscs, although only one class of the latter (gastropods) lives on land. In these animal phyla and in land plants, much of the diversification has involved anatomical characters, but groups such as fungi and nematodes (and some kinds of insects, notably aphids and weevil beetles) include hundreds of species that are very similar in structure, but differ biochemically. Nonetheless, biochemical properties may be as important as anatomical features in that they equip the organism to live under different conditions or to exploit different sources of food. However, because classification is based largely on anatomical characteristics, the organisms will tend to be classified together in the same categories.

Although one can suggest certain features, such as the presence of cuticle and limbs in arthropods or seeds in flowering plants, that seem to predispose a group to extensive adaptive radiation, it is usually very difficult to explain completely or precisely why certain groups are physiologically similar but have diversified anatomically and why others are structurally uniform but biochemically diverse, whereas others have undergone little adaptive radiation and so include only a few, apparently similar species. Species that belong to small taxonomic groups are not necessarily rare, inconspicuous or incapable of adapting to difficult environments: pogonophorans are large and abundant in one of the world's most hostile environments; in many areas of the arctic ocean, the plankton teems with sea-gooseberries (see Chapter 1, Section 1.6.2) and salps (see Section 2.2.1), and horsetails are a common weed in gardens and in sheltered, boggy areas throughout Britain. Amphioxus (Figure 2.10), lampreys (Figure A.45a), arrow-worms (Appendix, Table A.1) and horseshoe crabs (Figure A.36a) are also very abundant in certain places but have only a few living relatives (although there may be many extinct species in the group). Earlier this century, lampreys (one of a very few surviving kinds of primitive jawless vertebrates) so successfully colonized certain large lakes in North America that stocks of teleost fishes (one of the most recently evolved, and currently the most abundant and diverse group of vertebrates) were seriously depleted.

There are also some surprising 'failures' of adaptive radiation. For example, nearly a million species of insects have been described, with habits ranging from parasitism (e.g. lice, fleas) to long-distance migration (e.g. locusts), including numerous different freshwater forms (mayflies, dragonflies, water beetles, etc.). However, only about a hundred insect species, all of them quite small and many of them rare, live permanently in the sea, and many of these are confined to shores or live on floating driftwood or multicellular algae. So far as we can tell from the fossil record, insects have never been a significant component of the marine fauna. So, in spite of the fact that most other groups, including many kinds of arthropods, have marine forms (crustaceans are the most abundant multicellular organisms in many oceans), and several insect orders are abundant and diverse in freshwater, insects are almost absent from the three-quarters of the Earth's surface that are covered by oceans. Nonetheless, the fact that there are a few species of insects in the sea indicates that there are no fundamental physiological reasons why insects cannot survive in a marine environment.

◇ Referring to Chapter 1, can you name another case in which a few species of an abundant group have successfully colonized a habitat but not undergone adaptive radiation there?

◆ *Hydra*. Although nearly all living (and fossil) cnidarians (Chapter 1, Section 1.6.2) are marine, *Hydra* is widespread and often common in freshwater ponds and lakes.

It is also difficult to establish the processes that led to diversification and the circumstances under which they occurred. There are very few examples of living, or even fossil, species that are clearly intermediate between major taxonomic categories (see Section 2.2.3). Adaptive radiation can only be identified in retrospect: there is no scientific basis on which biologists can predict which kinds of organisms will become abundant and diverse several millions of years from now, or the course of evolutionary changes in the future.

Summary of Section 2.2

Larvae are structurally distinct from the adults of the same species and are adapted to different diets and habits and do not reproduce sexually. In some animals, the larval stage is brief, but other species spend almost the entire life cycle as larvae. Many invertebrates, fishes and amphibians have one or more larval stages in their development; there are two main kinds of life cycle in insects and several different kinds of larvae. Larvae of different groups may have similarities that are not apparent in the adult forms, and so may provide clues about their origin and interrelationships.

Although fragmentary and biased, the fossil record is the principal source of information about the evolutionary history of organisms. In a few cases, preservation is complete enough to provide insight into the structure and habits of extinct organisms and their possible interactions with other species. Some groups of organisms have diversified and given rise to numerous different species, while others, including some widespread and abundant forms, have undergone very little adaptive radiation.

Question 3 (*Objective 2.4*) Which of the following statements about larvae are normally *true*?

(a) The diet and habits of the larval stage are often different from those of the adults of the same species.

(b) Certain kinds of food can be exploited only by larvae.

(c) Larvae never produce gametes.

(d) The presence of a larval stage in the life cycle hastens the adaptive radiation of the species.

(e) The presence of a larval stage in the life cycle reduces the chance that the adult forms will become extinct.

Question 4 (*Objective 2.5*) Which of the following statements are *true* about the relevance of larval forms to determining relationships between groups of organisms?

(a) Larvae are always more similar than the adult stage to the ancestral form.

(b) During development, organisms always pass through stages that resemble their ancestors.

(c) Sometimes features of the basic body plan are more discernible in the larva than in the adult.

(d) Sometimes features of the basic body plan are more discernible in the adult than in the larva.

(e) The larvae of organisms that spend only a small fraction of their lives as adults are more similar to the ancestral form than those in which the larval stage is transient.

Question 5 (*Objective 2.6*) How reliably can the following kinds of information be obtained from the fossil remains of extinct organisms?

(a) The gross anatomical structure

(b) The habitat in which they lived

(c) The food that they used and the organisms that ate them

(d) The time of the first appearance of the species

(e) The time of extinction of the species

Question 6 (*Objective 2.7*) Which of the following kinds of animal have undergone extensive adaptive radiation?

Hydra	Cephalochordata
Peripatus	Carnivora
Lepidoptera	humans
Gastropoda	

2.3   EVOLUTION FROM SEA TO LAND

The evolution of terrestrial organisms from aquatic ancestors was one of the most important events in the history of the Earth. It took place in many different lineages and involved changes in many different aspects of structure, habits, physiological capacities and life cycles. Many of the concepts and sources of information outlined in Section 2.2 can be combined to reconstruct the major events in the colonization of land and the circumstances under which it took place.

2.3.1   The history of colonization of the land

Fossil evidence for life appeared first in marine sediments and, for the first 300 Ma of their evolutionary history, all organisms were probably confined to the oceans, salty lakes and estuaries. Most of the major groups of plants and animals originated in the sea. There are no fossils in rocks formed in terrestrial or freshwater environments before and during the Cambrian, so we assume that the land and its lakes and rivers were barren. By the Ordovician, remains of algae, primitive fishes, various kinds of worms and arthropods occurred in freshwater deposits, and during the Silurian, about 70 Ma later, the first land organisms appeared in the fossil record.

◇ Would you expect the fossil record of terrestrial organisms to be as complete as that of marine or freshwater species?

◆ No, fossils do not form as readily in terrestrial environments. Terrestrial organisms that fall into water are more likely to be preserved, but in such cases, the interpretation of anatomical characters provides the only evidence that they did in fact live on land.

Micro-organisms and plants, particularly non-woody plants, are poorly represented in the fossil record.

◇ For which of the animal phyla would you expect the fossil record to be most extensive, and why?

◆ Molluscs, arthropods and vertebrates are probably most abundant as fossils, because most members of those phyla have hard parts that would be likely to be fossilized.

Biased and incomplete though it almost certainly is, the fossil record is the only source of information we have about the timing of the first appearance and the relative abundance of terrestrial organisms. Look back at Figure 2.13 and Table 2.1. Several lineages of plants colonized the land during the Devonian and became abundant during the second half of the Palaeozoic and the early Mesozoic (Figure 2.13), forming dense, lush forests. Some Palaeozoic ferns and fern-like plants were the size of a large bush, and many gymnosperms now extinct grew to be tall trees, supported by tough, dense wood that is familiar to us in its fossilized form as coal. However, since the Cretaceous all the kinds of terrestrial plants except the angiosperms have decreased in numbers, though the true ferns and conifers declined less drastically than the others. The evolution of the angiosperms coincided with, and probably partly caused, the partial extinction of these ancient plants that had dominated the land for 200 Ma. Changes in life cycle and mechanisms for reproduction are among the most important adaptations that promoted the colonization of land. Seed-bearing plants first appeared in terrestrial habitats, and the great majority of species still live on land. Although a few (e.g. water-lilies, eel grass and rushes) have secondarily become aquatic, they have retained almost all of the major features of structure and life cycle of their terrestrial relatives.

2.3.2 Adaptations to living on land

The sensation of lightness that you feel when swimming underlines a crucial difference between water and air: water is much denser than air, and therefore the support available for living organisms is quite different in the two media. This contrast is obvious if you compare the floppy mats of seaweeds on the shore when the tide is out with the fronds floating vertically when they are immersed in water. It is true that many land organisms rest on 'solid' ground (or in burrows or on or in other biological structures such as trees), but even quite small bodies cannot stand erect without some kind of supporting skeleton. The most striking structural and physiological adaptations to living permanently on land are found in larger multicellular plants and animals.

◇ What is the principal structural tissue in large terrestrial plants?

◆ Wood, which appears in angiosperms and gymnosperms (Chapter 1, Section 1.4.3).

Another important difference between terrestrial and aquatic environments is the availability of the basic requirements of living organisms, food, water, oxygen, shelter and, in the case of plants, light, mineral nutrients and carbon dioxide for photosynthesis. The sea contains dissolved mineral salts and gases that can be absorbed over the whole surface of the body of small organisms such as bacteria, algae, protistans and small animals. But for terrestrial plants, these essentials are unevenly distributed between air and soil. Air contains a variable amount of water vapour, 20% of its volume is oxygen and 0.03% is carbon dioxide.

Soil consists of solid particles such as clay or sand and the decaying remains of dead plants and animals, all of which are normally coated with water and interspersed with small pockets of air. The water film contains dissolved mineral salts and is a habitat for minute soil organisms, particularly bacteria, fungi and protistans that feed on the decaying remains. Some larger animals, such as moles, earthworms and many kinds of small arthropods, spend their entire lives in the soil, feeding upon smaller soil organisms. However, green plants cannot live entirely underground because light does not penetrate through soil. Thus land plants need both roots that have access to soil, whence they obtain water and mineral nutrients, and shoots that protrude into the light above ground. The leaves and stems of terrestrial plants are impermeable to water over most of their surface. Respiratory and photosynthetic gases enter and leave the plant through specialized pores, called stomata, that can be closed when water is scarce.

The necessities of structural support, avoidance of desiccation and access to nutrients affect all adult organisms living on land; the immature stages of the life cycles face all these problems and some additional ones. The gametophyte stages of ferns and fern-like plants are aquatic and in these groups of plants and in mosses, movement of gametes and fertilization must take place in a watery environment (see Chapter 1, Section 1.4). So, although certain stages of these plants survive for long periods in a dry environment, they cannot complete their life cycle without water. The evolution of flowers (or cones) and the formation of seeds in seed plants means that the gametes form and fertilization takes place on the parent plant, thereby eliminating an aquatic stage in the life cycle. Because of their small size and immature tissues, the embryos and seedlings of seed plants are more vulnerable than adults to drying and to extremes of temperature, but they are nonetheless much more emancipated from dependence on water than other groups of terrestrial plants.

Animals living on land face many of the same problems as terrestrial plants: their bodies need mechanical support and are susceptible to damage from desiccation and from rapid and extensive changes in temperature. However, unlike plants, most animals can move about and can therefore hide in the shade or in burrows, or return to the water at certain seasons or times of day. Many terrestrial invertebrates are confined to moist, shaded habitats, such as soil, undergrowth or leaf litter, or are active only at night when the air is cooler and more humid. Like plants, the gametes and embryos are the stages of the animal's life cycle that are most affected by the absence of water. Modifications of the reproductive system that provide a suitable environment for these delicate stages were thus a major element in the adaptation of animals to living on land.

Many of the first terrestrial plant and animal fossils are found in rocks formed inland. They resemble freshwater forms more closely than marine species, suggesting that the land was first colonized by organisms whose immediate ancestors lived in rivers or estuaries.

◇ How could living in freshwater facilitate the evolution of adaptations to terrestrial life?

◆ Freshwater has a much lower concentration of salts than the sea and, because it is usually shallower, undergoes larger fluctuations in temperature than the sea. Adaptations of the excretory organs that maintain the normal concentration of salts in the body, and of metabolism that withstands extremes of temperature, are likely to be well-developed in freshwater organisms.

Stagnant ponds and lakes often contain little oxygen even when shallow, and the larger animals that inhabit them often have organs, such as lungs, that enable them to obtain oxygen directly from the air. Such features equip organisms for life on land better than adaptations to living in the turbulent, well-oxygenated surface waters of the sea. In fact, a freshwater stage is almost universal in the evolution of terrestrial organisms from marine ancestors; there are very few instances of intertidal organisms becoming fully terrestrial, even though their habitat requires them to tolerate periods of desiccation and extremes of temperature.

Like most modern terrestrial micro-organisms, and many small metazoans, the first colonists of the land lived in microhabitats that were essentially aquatic. For example, some bacteria and protistans colonize temporary aquatic habitats, such as puddles, but they form metabolically inert spores when the water dries up. Inhabitants of damp soil are protected from desiccation and extremes of temperature and many soil micro-organisms live in the films of water that coat the minute mineral particles. Completely dry soil cannot support life except in the form of dormant spores, and it quickly becomes dust or sand.

Animals first colonized the land in significant numbers during the Silurian.

◇ What would the Silurian terrestrial environment have been like?

◆ The information in Table 2.1 and Figure 2.13 indicates that, apart from lichens (which will be discussed in Section 2.4.1), the only terrestrial plants were small, leafless psilopsids. There would have been little shelter from strong sunshine and severe weather.

In other words, in the Silurian the terrestrial environment was even less hospitable for animals adapted to living in water than it is now. Nonetheless, terrestrial forms evolved from many different lineages of animals: platyhelminths, annelids, nematodes, molluscs, arthropods and vertebrates and several less familiar groups. The arthropods and vertebrates are the most thoroughly adapted to terrestrial life, so we will use these groups as examples of how adaptations to terrestrial life may have evolved.

Terrestrial arthropods

Like most other animal phyla, the earliest arthropods were marine and several lineages have been abundant in the sea from the Cambrian to the present day. Fossil arthropods are found in terrestrial deposits of Silurian age, making them the first group to colonize the land in significant numbers. The first terrestrial arthropods were small, elongated animals superficially similar to *Peripatus* (see Section 2.2.3), which probably lived in or on soil.

◇ What aspects of the structure of arthropods would make them particularly suited to life on land?

◆ Their cuticle and localized respiratory surfaces (see Chapter 1, Section 1.6.4).

Cuticle forms an impermeable outer covering for all stages of the life cycle from egg to adult, so the embryos and juveniles can be almost as well protected from desiccation as the larger adults.

◇ Which stage of the life cycle cannot be protected by cuticle?

◆ The gametes.

Fertilization normally involves an actively swimming sperm cell entering and fusing with the ovum and so requires an aquatic environment. In the majority of arthropods, including some aquatic forms such as certain crustaceans, fertilization takes place inside the female's body and, in most arthropods, the mother's oviduct secretes a tough, cuticular case around the zygote immediately after fertilization, thereby protecting the developing embryo from desiccation.

Arthropods from several different lineages live on land. Among the major living groups, the terrestrial crustaceans are the least well-adapted to withstand desiccation and extremes of temperature.

◇ What are the respiratory organs of typical crustaceans?

◆ In most crustaceans, parts of the limbs are modified to form gills, which are exposed to the surrounding water (see Chapter 1, Section 1.6.4).

The gills of the most familiar terrestrial crustacean, the woodlouse (Figure 2.15), are enclosed in flaps of impermeable cuticle, and very little gaseous exchange can take place unless the gills are moist. Woodlice dip their gills into

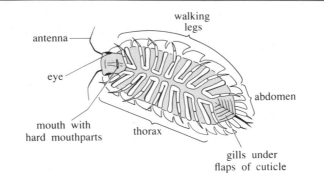

Figure 2.15 Woodlouse, a terrestrial crustacean (order Isopoda), seen from below.

OUTSIDE   INSIDE

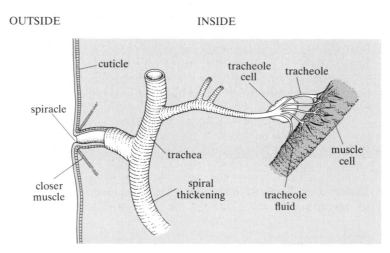

Figure 2.16 Arthropod trachea, partly cut away to show the internal structure. The cuticle-lined trachea is relatively impermeable. Gas exchange takes place through the unlined tracheoles.

drops of water if they are confined in warm, dry air for longer than a few minutes and many species die from desiccation within hours if denied access to freshwater. Most woodlice are thus confined to damp, shaded habitats such as dense undergrowth and rotting wood.

Among the earliest land arthropods, and the most completely adapted to terrestrial habitats, are the insects and arachnids (spiders, mites, scorpions and harvestmen; Chapter 1, Section 1.6.4). All the larger terrestrial forms have a system of air-filled tubes, called tracheae (Figure 2.16) that extend to all tissues of the body from small openings at the surface and end in fine branches lined with permeable tissue, through which gaseous exchange takes place. The external openings of the tracheae are called spiracles, and in insects they are usually arranged in one or more pairs on each segment. The lining of the tracheoles is permeable to water as well as to gases, so the respiratory surfaces are a potential route for water loss. In arthropods adapted to dry climates or prolonged exposure, the spiracles are closed completely by flaps of cuticle operated by muscles, except when strenuous exercise, such as flying, requires all respiratory surfaces to be operational. Tracheae are also present in some secondarily aquatic arthropods such as freshwater insects and insect larvae (see Section 2.2.1) but the principal respiratory organs of most primarily aquatic groups such as crustaceans are gills.

The excretory system of insects is not only efficient at conserving water, but is also able to compensate for imbalances in the diet by selective excretion or retention of essential nutrients.

Insects are, and have been since the Carboniferous, by far the most abundant and diverse organisms on land and are also common in freshwater, but they are a very minor component of the marine fauna (see Section 2.2.4). A very few spiders live in still, shallow water such as bogs or ponds, but there are no truly marine species.

Terrestrial vertebrates

Figure 2.17 shows the first appearance in the fossil record and possible interrelationships of the major categories of living vertebrates. You are already familiar (Chapter 1, Section 1.6.7) with the essential features of the two most abundant kinds of fishes in modern waters, the chondrichthyans and the teleosts. Since the Cretaceous, the great majority of osteichthyans have been teleosts. However, neither of these major groups of fishes is thought to be ancestral to the tetrapod vertebrates. Terrestrial vertebrates are believed to have arisen from the fleshy-finned bony fishes, the Sarcopterygii. Sarcopterygian fishes are an ancient group, appearing first in the early Devonian, but they have never been dominant members of the fish fauna. There are four living genera, three of them freshwater lungfishes (Figures 2.18a and b) in Australia, tropical Africa and South America, and, of a rather different ancestry from the rest, a single marine genus, the coelacanth, *Latimeria* (Figure 2.19a). Although quite common as fossils in the late Palaeozoic and Mesozoic, marine sarcopterygians were believed to have been extinct for more than 50 Ma until a specimen was caught off the coast of South Africa in 1938. The living species bears remarkable resemblance to the extinct forms of Cretaceous age (Figure 2.19b). *Latimeria* was not seen alive in its natural habitat until the mid-1980s, when biologists equipped with deep-water submarines explored the deep sea off the Comoro Islands in the Indian Ocean.

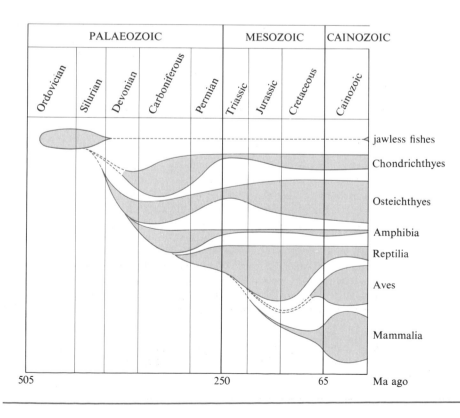

Figure 2.17 The first appearance, relative abundance and possible evolutionary relationships between the major classes of vertebrates.

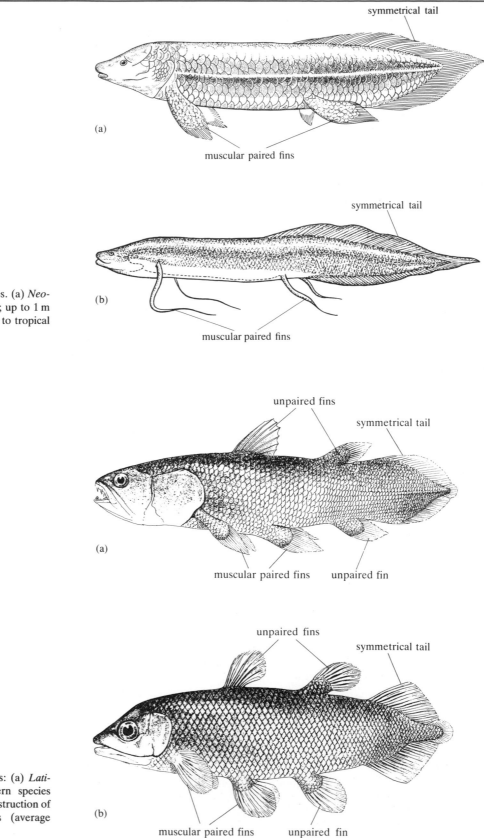

Figure 2.18 Modern lungfishes. (a) *Neo-ceratodus* (native to Australia; up to 1 m long); (b) *Protopterus* (native to tropical Africa; up to 1.5 m long).

Figure 2.19 Coelacanth fishes: (a) *Lati-meria chalumnae*, the modern species (about 1.5 m long); (b) Reconstruction of a Cretaceous fossil species (average length: 0.5 m).

Lungfishes all live in freshwater, often in swamps, where plants and smaller animals quickly use up almost all the oxygen in the warm, stagnant water. As their name implies, lungfishes have lungs as well as gills. The lungs are important as a respiratory surface, and also act as a buoyancy organ. Indeed, the African species *Protopterus* and the South American species *Lepidosiren* drown if denied access to the air. Large *Protopterus* can live for several days out of water in damp surroundings and survive periods of drought by burrowing in mud and secreting a cocoon around themselves. Adaptations to living out of the water are best developed in adult lungfishes; mating, incubation of the eggs and maturation of the juvenile stages all take place in lakes or swamps, so the life cycle cannot be completed away from water.

◇ Which other organisms are terrestrial when sexually mature but require water to complete their life cycle?

◆ Several groups of non-seed-bearing plants, particularly bryophytes and ferns (see Chapter 1, Section 1.4). Several groups of insects including dragonflies, mayflies and mosquitoes and many amphibians (Chapter 1, Section 1.6.7) such as frogs and newts are terrestrial as adults but aquatic as larvae.

Terrestrial vertebrates should be recognizable as such in the fossil record by having paired limbs instead of fins, lungs instead of gills and a waterproof outer covering. By the middle of the Carboniferous, there were several kinds of fairly large vertebrates that had stout limbs and were clearly able to spend at least part of their adult lives on land. Such animals could properly be classified as amphibians.

◇ Can most modern amphibians complete their life cycle without water?

◆ No. Most frogs and newts mate in water and have aquatic larvae (e.g. tadpoles).

◇ Is it possible to find out about how fertilization and larval development took place in extinct animals?

◆ Generally not. Many larval arthropods (see Section 2.2.1) have hard cuticle but the immature stages of molluscs, echinoderms and aquatic chordates have few or no hard parts and are rarely preserved as fossils.

In fact, it is usually impossible to determine with any certainty whether a specimen is an adult of a small species, or a juvenile of a large one.

One of the most important features that enabled vertebrates to complete their entire life cycle on land was the amniote egg (Figure 2.20); such eggs are relatively large and contain a high proportion of yolk from which the embryo obtains nutrients during its development. Normally the embryo begins to develop as soon as the egg is fully formed and development generally proceeds faster at higher temperatures. Incubation lasts from less than two weeks in some small birds to many months in crocodiles, snakes and tortoises. The tough shell is permeable to oxygen and other gases, and gaseous exchange takes place across the allantois. The shell also provides protection from desiccation and to some extent from predation. However, the small pores in the shell can become waterlogged, and the embryo may drown if the egg is completely immersed in water for longer than a few minutes. A few species of snakes and lizards are viviparous: the egg shell is greatly reduced and modified and the eggs are retained inside the female's body, where the embryos complete their development and are born as free-living juveniles.

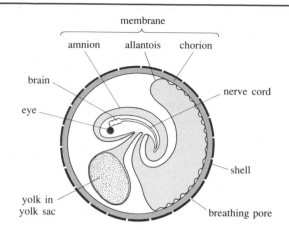

Figure 2.20 Diagram of an amniote egg to show the developing embryo in the embryonic membranes.

All living reptiles, birds and monotreme mammals (see Chapter 1, Section 1.6.7) have amniote eggs and the remote ancestors of marsupial and eutherian mammals may also have had them. However, it has proved very difficult to establish which fossil forms were the closest ancestors of the lineage in which the amniote egg first appeared, because many of the terrestrial features of the adults, including lungs and muscular limbs, occur in ancient (and living) fishes adapted to living in swamps and stagnant pools. The essential difference between fins and limbs is that limbs terminate in a foot (or hand) with toes or fingers, but a fin carries a web supported by fin-rays. The skeleton of the hand or foot is quite different from fin-rays in origin and in arrangement and there are no known intermediates between fins and limbs.

Vertebrates have been abundant and diverse on land ever since amphibians first appeared in the Carboniferous. Although several kinds of reptiles (e.g. sea snakes, terrapins, turtles), birds (e.g. ducks, penguins) and mammals (e.g. whales, seals, sea cows) live in water, these secondarily aquatic forms have not displaced the fishes as the most abundant and diverse groups of vertebrates.

Summary of Section 2.3

Although life probably began in the sea, many different groups of plants and animals include terrestrial forms. The adaptations necessary for living on land, such as support, protection from desiccation and the ability to breathe air have evolved in different ways in the different groups.

Question 7 (*Objective 2.8*) Which of the following features occur *only* in terrestrial species?

Skin

Hard skeleton

Lungs

Tracheae

Limbs

Jaws

Wood

Flowers

Leaves

2.4 WAYS OF LIVING

The colonization of land increased the diversity of habitats and so led to a greater variety of organisms. Differences in diet and mechanism of feeding are some of the many factors that promote diversity of species. Most familiar animals and plants are free-living: they can obtain food, shelter and other essentials in several alternative ways, and they flourish in association with a variety of other species. For other organisms the relationship with the environment is much more specific: they frequently or invariably occur in intimate association with one or a narrow range of other species and only thrive in such associations. Such interdependent relationships between organisms of different species are called **symbiosis** ('living together').

In some cases, biologists can identify ways in which *both* partners benefit from living together: they may provide each other with food or protection from predators. Such a symbiotic relationship is called **mutualism**. A symbiont that obtains much or all of its food or protection from the other organism, and in so doing, depletes its partner's food supply or damages its tissues, is called a **parasite** and the organism that is parasitized is called the **host**. However, many other symbiotic relationships are more complicated and the degree of interdependence, and what benefits the partners derive from it, may be unclear, making it impossible to be sure whether it should be described as mutualistic or parasitic. In some cases, symbiotic relationships that were formerly believed to be parasitic turn out, on more thorough investigation, to be cases of mutualism.

Often the distinction between free-living habits and symbiosis is not clear cut; thus some insects, e.g. many caterpillars and weevils, can digest only a few kinds of foliage or seeds and are found only on one or a few closely related species of plant. These 'free-living' species are as uniquely dependent on just one other species for food and shelter as many symbionts are on their hosts. Conversely, some 'symbionts' prosper on a variety of species, or may sometimes live independently. Many species are symbiotic for only part of their lives (e.g. as a larva) and may be free-living, or a symbiont of a different kind of organism at other stages of their life cycle.

2.4.1 Mutualism

Often, though not invariably, mutualism is nutritional: each partner obtains or synthesizes particular nutrients that it 'swaps' with its partner, thereby augmenting, or more often diversifying, the diets of both of them. Sometimes the relationship becomes so intimate and elaborate that the partners are never found independently in the wild, although it may be possible to culture them alone in the laboratory under optimized conditions.

Although often spoken of as though they were a single organism, **lichens** (Plate 3e and Figure 2.21) are an intimate and physiologically elaborate mutualism between a fungus, usually an ascomycete (Chapter 1, Section 1.5)

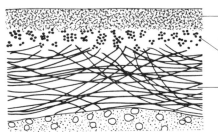

upper cortex
(fungal), 10–15 μm

algal layer, 10–15 μm

fungal medulla
500 μm or more

Figure 2.21 Internal structure of a simple rock-encrusting lichen. The algae live embedded in the mycelium of the fungal hyphae.

and an alga, usually a filamentous or unicellular chlorophytan alga or a cyanobacterium (Chapter 1, Section 1.2.2). Some lichens consist of an intimate but unstructured mixture of fungal hyphae and algal filaments, but there are also more elaborate forms in which the algae or cyanobacteria are concentrated in a layer, with the fungi outside and below them (Figure 2.21).

◇ How could a close association with an alga benefit a fungus?

◆ Algae photosynthesize in sunlight and synthesize carbon compounds that the fungus could utilize.

The benefit of the association to the alga is less obvious, but the ability of lichens to take up ions and salts very readily may be due mainly to the fungus. Lichens that contain cyanobacteria can fix nitrogen and hence can live on bare rocks and other places where organic nutrients are absent. In many lichens, the association continues throughout the life cycle: they 'reproduce' by spores that contain both fungal and algal cells (although in some forms, the fungus and the alga produce separate reproductive and dispersal structures). Some of the algae found in lichens also occur independently but very few of the 300 or so genera of fungi that form lichens are ever free-living. This mutualistic relationship is extraordinarily effective; although lichens grow very slowly, they are found in a great variety of habitats and flourish in some of the most severe climates and nutrient-poor environments in the world, such as exposed rock in high mountains, newly erupted volcanic lava and in arctic tundra, habitats in which few other kinds of organisms can survive. This mutualism is a very ancient way of living; some of the oldest fossil remains are of lichens.

Many protistans, including some ciliates such as *Euplotes* (Figure A.17b) and *Kentrophoros*, also form nutritional mutualisms with certain kinds of green algae, usually *Chlorella* species. The algae are enclosed within vacuoles and some of the larger protistans harbour scores, even hundreds, of symbionts. The protistans secrete simple nitrogen containing nutrients into the vacuoles, and take up sugars synthesized by algal photosynthesis.

◇ Where would you expect such organisms to live?

◆ In places exposed to bright sunlight. Such green protistans are abundant in the surface waters of lakes and in the oceans.

Many micro-organisms and fungi are capable of biochemical processes that do not occur in higher organisms. A few specialized species of bacteria and cyanobacteria are the only organisms that can reduce atmospheric nitrogen (in the form of N_2) to ammonia, which can then be incorporated into organic compounds such as proteins (see Chapter 1, Sections 1.2.1 and 1.2.2). The reasons for this somewhat surprising situation are still not clear, but it is the basis for some of the oldest and most biochemically elaborate and ecologically important mutualistic symbioses. Some angiosperms, particularly trees and herbs in the family Leguminosae, have nodules on the roots that harbour nitrogen-fixing bacteria. Some of the cyanobacteria that utilize atmospheric nitrogen form intimate and prolonged associations with several primitive plants, including ferns, liverworts and cycads. The nitrogenous compounds generated by these symbionts are taken up by the plant, thereby supplementing the nutrients available from the soil. Many leguminous food plants such as peas, beans and lentils owe their ability to produce large quantities of protein-rich seeds from poor soils to the presence of nodules containing nitrogen-fixing bacteria.

The root systems of many flowering plants, notably those of heathers, orchids and many trees such as beech and pine, frequently have fungal hyphae associated with them, forming **mycorrhizas**. Sometimes the fungus just lives on the roots, but often the relationship is anatomically intimate, with hyphae penetrating the root cells. Studies using radioactively labelled molecules show that carbon compounds pass from the plant to the fungus. Experimental studies also show that the angiosperm grows less well when deprived of its fungi. Until a few years ago, it was not clear exactly how the mycorrhizas contribute to the plant's wellbeing but recent studies suggest that the fungus facilitates uptake of mineral nutrients, and perhaps also affords some protection from plant diseases.

Mutualistic symbioses between animals and bacteria utilize the versatility of bacterial biochemical machinery for breaking down and synthesizing organic compounds.

◇ Which mutualistic relationship between an animal and bacteria has already been described?

◆ Pogonophorans derive nourishment from symbiotic bacteria in the trophosome (see Section 2.1).

Although many animals feed on higher plants, very few synthesize enzymes that thoroughly digest cellulose, the main component of leaves, wood and other tough plant materials. However, many species of fungi and bacteria can synthesize enzymes that break down cellulose. **Ruminant** mammals (deer, cattle, sheep, camels, etc.) harbour huge numbers of micro-organisms in part of the stomach and can therefore thrive on plants (particularly grasses) that other herbivorous mammals (e.g. humans) cannot digest. The symbionts break down the plant food to small, readily absorbed molecules and synthesize some nutrients, notably lipids, that are essential to the animal but absent from its diet. As the micro-organisms proliferate on the rich food supply, some are transported further along the gut, where they are digested.

Many other anatomically more complex but biochemically impoverished animals, including insects, molluscs and vertebrates, including humans and almost all other mammals, harbour large numbers of micro-organisms, mainly bacteria and protistans, in their guts. As well as releasing nutrients from foods that would otherwise remain undigested, the symbionts supplement the animals' biosynthetic processes by synthesizing certain vitamins and other chemically elaborate nutrients.

Termites are social insects that form large colonies that can destroy massive old, dry timbers impressively quickly. Their ability to digest wood is due to the presence in the gut of protistans of a kind not found elsewhere, either as a symbiont or free-living. By liberating nutrients that would otherwise be inaccessible, the protistans clearly benefit the host organisms, which starve to death if deprived of their symbionts.

Corals

Corals are cnidarians (see Chapter 1, Section 1.6.2) that form a hard skeleton. The great majority of corals are colonies of polyps that form by asexual budding from a single parent animal, and can grow to form elaborately branched structures (see Figure A.20). There are two main groups, the soft corals (Alcyonaria) in which hard spicules form in the mesogloea, and stony corals (Madreporaria), in which the skeleton is mostly calcium carbonate and forms below the ectoderm of the polyp (see Figure A.20a).

Like other cnidarians, all soft corals and some stony corals feed by trapping small organisms and detritus in their tentacles. Other stony corals, including some of the largest and most abundant forms, have reduced tentacles, mouth and enteron. For a long time, biologists were uncertain about how these corals obtained nourishment. The first clue was the presence of greenish-yellow cells between, and in some cases in, the cells of the coral. Further research showed that these cells were dinoflagellates (see Chapter 1, Section 1.3.2) which would live independently after separation from the cnidarian tissue if provided with light and suitable nutrients. These corals can also live for some time without the algae, and indeed, if kept in darkness, they sometimes expel their algae.

Using isotopically labelled precursors, biologists found that carbon atoms from carbon dioxide dissolved in seawater were incorporated in substantial quantities into glucose and some proteins in the cnidarian tissues in bright sunlight.

◇ In what biochemical process is external carbon dioxide incorporated into glucose?

◆ Photosynthesis. Animals cannot synthesize glucose from inorganic precursors. Some of the products of photosynthesis in the algae must have passed into the corals.

This result could simply show that the corals had digested the algae. However, the algal population remains constant over long periods of time, indicating that the rate at which the symbionts are destroyed is adjusted to their rate of proliferation (as in the case of symbiotic micro-organisms in ruminants). If the algae were removed, or if the symbionts were kept in darkness, the corals did not take up carbon dioxide from the seawater. Thus the corals and the algae are shown to have an intimate mutualistic relationship: substances generated by photosynthesis in the algal cells are an important, and in some cases the sole, source of nutrients for the corals, although many species also feed on tiny animals at night. Furthermore, the carbon dioxide removed by photosynthesis in the algae from the seawater surrounding the coral favours the precipitation of calcium carbonate, so aiding the formation of the skeleton of the stony coral. In bright sunlight, natural seawater and with a full complement of healthy algae, coral reefs can grow at up to $0.5\ \mu m\,h^{-1}$. Such corals can only flourish in clear, warm waters, where sunlight penetrates well, not because these conditions are necessary for the cnidarian tissues, but because bright light and well-aerated water are essential to support their symbiotic algae.

2.4.2 Parasitism

Parasites are symbionts that derive nutrients or some other benefit from their host at some cost to its wellbeing, but make no identifiable contribution to its ecology or metabolism. The presence of parasites is always in some way deleterious to their host, but in many, probably most cases, the harmful effects are trivial. **Ectoparasites** live on the host's skin, cuticle or outer surface; **endoparasites** live inside the host's cells or tissues or in its guts.

Fungi commonly known as 'rusts' and 'smuts' are parasites on green plants, including many economically important species such as cereals, peach trees and currant bushes. The potato blight that caused the famine in Ireland in the 1840s and Dutch elm disease that destroyed millions of elm trees in the 1970s are caused by parasitic fungi. A few fungi are also parasitic on animals, including humans; most produce only minor but irritating disorders such as athlete's foot and 'ringworm', but a few species can establish themselves in the lungs, causing serious illness.

There are relatively few completely parasitic plants, although more than 1% of all species of flowering plants obtain water and nutrients from the sap of other plants; the parasitic plant forms a special structure, the haustorium, that 'taps into' the host's tissues.

◇ Suggest an easily recognized feature that you would expect in all completely parasitic Chlorophyta.

◆ If they derived all their nutrients from the host, they would not photosynthesize and so would probably not be green.

A much larger number of plants are semi-parasitic, obtaining water and inorganic nutrients through roots that penetrate the host's stems and leaves, but retain the capacity for photosynthesis (e.g. mistletoe, some tropical orchids).

Many animal phyla include some parasitic species. Many leeches (Annelida) (see Chapter 1, Section 1.6.3 and Figure A.30) are ectoparasites on fish and other vertebrates. Several kinds of crustaceans (see Figures A.33, A.34 and A.35) and insects (Arthropoda) are ectoparasites for part or all of their life cycle, and a few are endoparasites. The phyla Platyhelminthes and Nematoda (see Chapter 1, Section 1.6.3) include large numbers of parasites, many of which are highly specialized endoparasites. The majority of animals and flowering plants, some authorities believe almost all, are parasitized by at least one nematode species, but parasitic platyhelminths are found mainly in molluscs and vertebrates, particularly aquatic species. Many such parasites have complicated life cycles involving several different larval stages, and elaborate structural and physiological adaptations.

The platyhelminth worm *Schistosoma* (class Trematoda) is an endoparasite of humans and certain freshwater snails. Its complicated life history is shown in Figure 2.22 and Plate 10b–f. The adult worms live attached by their muscular suckers to the walls of the veins surrounding the human gut or bladder. The male is flat but the sides of the body roll ventrally to form a groove in which the smaller, more rounded female is held. The pair mate, probably permanently, and thereafter the female produces fertilized eggs at a rate of 100–200 per day. The eggs are oval and have a single sharp spine and move around in the host's bloodstream until they lodge in small blood vessels of soft tissues such as lungs. The bleeding and other tissue damage caused by the adult worms and the eggs give rise to the debilitating disease, schistosomiasis (also called bilharzia), which afflicts up to 80% of the human population in parts of tropical Africa and Asia. Some of the eggs (possibly as few as 10% of the total) eventually enter the gut or bladder, and are eliminated with the faeces or urine. On contact with freshwater, the egg hatches to release a ciliated larva called a miracidium which can survive only 24 hours as a free-living, actively swimming organism. If it finds its way into the gut of one of several species of aquatic and semi-aquatic pulmonate snails, the miracidium sheds its cilia and bores through the wall of the host's gut, where it becomes a sporocyst. The sporocyst grows, feeding on the snail's tissues, and cercaria larvae form inside it, sometimes in large numbers. The cercariae are released from the sporocyst and leave the snail, which may die from a heavy infestation of parasites. The cercariae can swim actively in freshwater for several days, but they do not feed. When they encounter a human, they enter

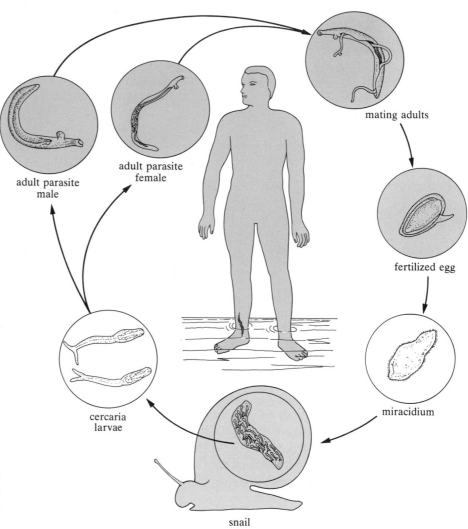

mating adults

fertilized egg

miracidium

snail

cercaria larvae

adult parasite male

adult parasite female

Figure 2.22 Life cycle of *Schistosoma mansoni*. Free-living larvae: white; larval stages parasitic in snails: grey; larval stages parasitic in humans: pink.

the bloodstream (usually through a small wound or sore) and, after a brief sojourn in the lungs, migrate to the liver and intestinal veins, where they feed on nutrients in the host's blood and grow to become adult worms.

◇ List some anatomical differences between adult *Schistosoma* and adult free-living platyhelminths (see Chapter 1, Section 1.6.3).

◆ The adult parasites lack eyes and there are specialized structures (suckers) that enable them to live in a continuously flowing bloodstream. The sexes remain together permanently, thereby promoting the continual production of fertilized eggs.

Such modifications or reductions of sense organs, nervous system and locomotory structures are typical of parasites, particularly endoparasites. However, other features of the anatomy, particularly reproductive structures, are highly complex and many are not found in free-living organisms. Although eyes would be useless for an internal parasite, other sense organs (e.g. those that are sensitive to chemicals) may be well-developed. In many other parasitic worms (e.g. tapeworms), the gut is reduced or absent; nutrients from the host are absorbed through the outer surface of the body, which has many of the properties of the inner lining of the gut of free-living animals. Sometimes such structural modifications are so drastic that it is difficult to identify the original body plan of a parasitic species, and hence to assign it to the right taxonomic category.

◇ What roles do larval and adult stages play in the life cycle of *Schistosoma* (see Section 2.2.1)?

◆ The larvae are free-swimming for a short time, but they do not feed while free-living. The sporocyst larvae produce cercariae asexually. The larvae parasitize a different host from the adults. Although adulthood is by far the longest phase, the adults are almost completely sedentary and do little more than mate and produce eggs.

The two brief free-living stages disperse the species more widely, and the secondary host (the snail) provides a valuable, if transient, supply of food that is often sufficient to support multiplication of the larvae. The human host provides the parasite with a secure long-term habitat and a nutritious food supply. Thus larval stages enable the species to live successfully in two or more very different habitats and to have 'the best of both worlds'.

◇ What is the difference between multiplication of the parasite in the human and in the snail?

◆ Reproduction is sexual in the human host, but larvae, by definition, are not sexually mature. Cercariae develop asexually in the sporocyst larva while it is in the snail.

Asexual multiplication of larvae means that a single infected snail can cause schistosomiasis in many humans. Once established in a human, the adult worms continue to release eggs for a long time, possibly the entire lifetime of the host, so a single human can infect many snails.

◇ What would you expect to happen to a miracidium or cercaria that was eaten by a fish or a shrimp?

◆ It would be digested in the same way as any other small organism.

This is probably the fate of most of the larvae, but in the gut of its snail host the miracidium escapes digestion, possibly by being impervious to the host's enzymes, or by producing a substance that inactivates those digestive enzymes.

In many animals, particularly mammals, the presence of any foreign biological material (and many non-biological materials) provokes an **immune response**, in which white blood cells surround and destroy the foreign substance or organism.

◇ Would you expect *Schistosoma* to provoke an immune response in its human host?

◆ Yes, the adults live in veins and are surrounded by the host's blood.

Adult *Schistosoma* injected into a fish or a bird are quickly engulfed by the white blood cells, but somehow evade the human immune system. One way in which parasites 'hide' from the host's immune system is by adsorbing the host's own proteins onto their surface, thereby producing a form of 'molecular camouflage'. Avoiding destruction by digestion or by the immune system is one of the most important and intricate adaptations of parasites to their way of life and may be a major reason why many parasites thrive only in one or a narrow range of host species. Such 'host-specificity' gives rise to large numbers of similar-looking parasite species that differ from each other in anatomically slight but functionally important ways.

◇ Which of the two hosts is likely to be most effective in dispersing the parasite?

◆ Humans; they migrate long distances and their sewage treatment system (or the lack of it) results in their excrement entering rivers and ponds.

◇ How could schistosomiasis be prevented?

◆ By eliminating all the adult worms from humans; by preventing human sewage from entering freshwaters; by destroying all snails (or all humans) in which the parasite can thrive; by making all freshwater toxic to the larvae; by keeping humans (or snails) away from infected water.

None of these options is totally feasible. The last two options work fairly well, at least on a short-term basis and where humans have access to purified water, and in some areas the numbers of snails have been significantly reduced by introducing fish or ducks that eat them, or by poisoning them with chemicals.

Parasites such as *Schistosoma* damage their host by poisoning, injuring or exploiting its tissues. A heavy infestation of such parasites can severely weaken or kill the host. Such effects tend to bring the parasites to the attention of biologists, particularly if humans or domestic livestock are involved. Sometimes the host's physiological responses to the parasite, such as vomiting or scratching, actually facilitates the maturation or transmission of the parasite.

◇ Will causing diarrhoea in its human host be to the parasite's advantage?

◆ Yes. Unless the person stays in the same place all the time, more pools will be infected more frequently with hatching eggs.

However, parasitism is not necessarily, or even usually harmful, and many parasites have few detectable effects on the well-being of their host.

◇ What will happen to the parasite if its host becomes ill or dies?

◆ If the host dies, the parasite's food supply and mobility cease and it will die too unless it can move to another host. When the life cycle is complex, only certain stages are capable of migration from host to host. For example, the adult stages of *Schistosoma* are unable to move from host to host.

All parasites derive some benefit from the host and many cannot complete their life cycle at all except in or on a living host. A parasite that drastically weakens or kills its host reduces its own chance of breeding successfully.

Ectoparasites

The Anoplura is one of several insect orders that consists largely or entirely of ectoparasites. Commonly known as lice, they are wingless throughout their lives, although features of their adult anatomy and larval development strongly suggest that they evolved from winged ancestors. Anopluran lice are found only on mammals, and both larvae and adults live in hair and suck blood through sharp, piercing mouthparts that puncture tiny holes in the host's skin. Most lice can crawl only slowly on their weak legs and the adults die of cold if kept below 25 °C, or of starvation if removed from the host for longer than a few hours. Many generations can multiply in the fur of a single mammal, and are transmitted from host to host by body contact between the sexes, or between mother and offspring.

◇ What features of the anatomy of anopluran lice (Figure 2.23) could be adaptations to ectoparasitism?

◆ The body is squat and flat, and the legs terminate in hooked claws, suitable for clinging onto hairs. The eyes and antennae are relatively small compared to those of free-living insects (see Figures 2.8 and 2.24 and Chapter 1, Section 1.6.4), suggesting that these sensory capacities are reduced.

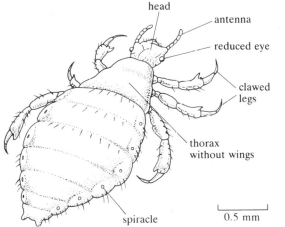

Figure 2.23 Human body louse, *Pediculus humanus*.

The structure of the claws of many species of lice is suitable only for clinging to hair of a certain thickness and texture. Lice that infest an inappropriate species may fall off or be removed easily by grooming. Many ectoparasites are thus specifically adapted to thrive only on one or a narrow range of hosts. Most, possibly all, species of terrestrial mammals (except rhinoceroses, hippopotamuses and other almost hairless forms) are parasitized by at least one kind of anopluran.

◇ On what parts of the human body would you expect anopluran lice to breed?

◆ Mainly on the head, where there is dense 'fur' similar to that in which lice live on most other mammals. Most of the rest of the body is almost hairless and hence unsuitable for lice.

The human head louse *Pediculus humanus capitis* is indeed restricted to the scalp and hair, but a subspecies, the body louse, breeds in clothing and in many parts of the world is the commonest ectoparasite of humans.

◇ From these facts, what can you conclude about the evolutionary history of the human body louse?

◆ The body louse must have acquired the ability to breed in clothing in the relatively short time (perhaps less than 10 000 years) since humans began wearing woven cloth.

Lice do little harm themselves beyond causing minor bleeding and irritation, but they transmit their own parasites, the rickettsiae (see Chapter 1, Section

1.2.1) that cause typhus fever, to their human hosts. Within hours of ingesting blood from a person suffering from typhus, the rickettsiae invade the cells of the insect's gut and multiply there, causing the cells to rupture. The rickettsiae released into the gut contaminate the insect's faeces. Although the louse dies within a few hours of infection, the rickettsiae remain virulent for several months, and can infect other humans if these faeces, or the bodies of the dead lice, are inhaled or enter through small wounds.

Although some species of lice are generalists, parasitizing several different species of host, many others are found on only one species, and in some cases are confined to only one part of the body (e.g. head lice, body lice). Such species-specific association between parasites and their hosts may last for a very long time. Studies of the resemblances between bird lice (order Mallophaga) have aided the elucidation of evolutionary relationships between hosts. For example, flamingos (Chapter 1, Section 1.6.7, Figure 1.35) are tropical, freshwater birds that feed on diatoms (Chapter 1, Section 1.3.2) using a unique filter mechanism in their beak. Taxonomists first classified them as related to another group of long-legged wading birds, the storks. But a survey of the lice on freshwater birds revealed that those parasitizing flamingos were more similar to those on ducks than to those on storks, supporting the conclusion that flamingos evolved from duck-like ancestors.

2.4.3 Interactions between free-living organisms

Species-specific interactions involving elaborate adaptations of structure or behaviour also occur among free-living organisms. In many cases, such relationships are as intricate as the relationships between parasites and their hosts described in Section 2.4.2, and play an equally important role in promoting species diversity.

◇ Why would the adaptations of parasites to their host be more readily recognized as examples of symbiosis between free-living organisms?

◆ Because parasites and hosts form prolonged, physical contact with each other, the relationship between the organisms is obvious. However, museum specimens of free-living organisms are usually collected without their associated organisms and a biologist examining an isolated plant or animal may not be familiar with its habits in the wild.

Relationships that do not involve prolonged physical proximity may be more difficult for laboratory-based biologists to identify but they nonetheless promote diversity and provide functional explanations for many otherwise inexplicable features. The roles of animals as **pollinators**, herbivores and dispersal agents of angiosperm plants provide many examples of specific and structurally elaborate interactions between free-living organisms.

The pollen grains that contain the gamete in gymnosperm and angiosperm plants are enclosed in a tough case and are not motile. The female gamete, the ovule, is sessile and fertilization takes place on the parent plant (Chapter 1, Section 1.4.3).

◇ How does this system compare with the mechanism of fertilization in typical animals?

◆ The sperm of almost all animals is motile, and in many species, the male actively moves towards the female. In most animals, fertilization takes place in or near the female after the male releases his gametes nearby, or introduces them into her body.

Thus the system of reproduction of higher plants presents the problem of transporting the pollen. The pollen of almost all gymnosperms (and of some angiosperms, e.g. grasses) is dispersed by the wind, but that of many other flowering plants is transported by animals, including birds, bats and various arthropods, particularly insects. Most vertebrate pollinators are tropical and subtropical; in temperate regions of the world, flying insects, particularly hymenopterans (bees), lepidopterans (butterflies and moths), dipterans (flies) and coleopterans (beetles) are the principal pollinators. The pollinators are attracted to flowers, often because they suck the nectar, pollen or other component of the plant, and, by one means or another, some pollen becomes attached to the body. Pollen acquired on a visit to one flower may rub off the animal while it is on another flower, thereby transporting the pollen between different plants and promoting cross-fertilization.

Bees are important as pollinators for a wide variety of plants, including some economically important species such as clover, alfalfa and many fruit trees. Figure 2.24 shows some features of typical bees: the third pair of legs has a dense mass of cuticular bristles to which the pollen sticks firmly, sometimes as many as 15 000 pollen grains on each bee. The long flexible mouthparts form a tongue through which the bee sucks up nectar. The eyes are large and experiments show that honey-bees have good vision and sense of smell, and readily learn to find even widely-scattered flowers. They are fast, powerful fliers, some species covering up to 23 km in a single day; although they do not fly during rain, some species, especially bumblebees, continue foraging in cool weather.

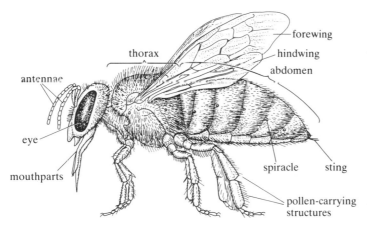

Figure 2.24 External features of a honey-bee.

Nectar has no known role in the physiology of plants and its sole function seems to be to promote the transfer of pollen between plants by attracting pollinators into the flowers. However, many other animals that do not transport pollen effectively find nectar an appetizing and nutritious food (including humans, who eat it as honey). In typical bee-pollinated flowers (Plates 7d and e), the nectary is at the base of the flower and, in reaching it, a bee of the appropriate size and shape 'pushes past' both the anther and the stigma, thereby collecting fresh pollen and depositing pollen from a previously visited flower.

Larval bees are legless grubs that, in many species, are fed by the adults until they reach the adult stage.

◇ Would nectar, which is almost pure sugar and water, be an adequate diet for a growing insect?

◆ No. A source of proteins and lipids is essential for growth.

The main source of these essential nutrients is pollen. Only a fraction of the pollen collected by bees is conveyed from one flower to another; the rest is fed to the larvae.

◇ Would bees be less likely to enter flowers in which their bodies would not touch the stigma or anther?

◆ Yes; although the bees might be able to reach the nectar, they would not obtain any pollen.

In other words, bees that do not act as efficient pollinators also do not obtain nutritionally adequate food for their larvae.

◇ Would animals that visited many different kinds of flowers be effective pollinators?

◆ No. The pollen from different species would get mixed up. Successful fertilization takes place only if the ovum and pollen are from the same species.

An important function of flowers is to attract the appropriate pollinators and deter those that, by taking nectar without transporting pollen, act as herbivores; much of the immense range of species-specific colours, shapes, patterns and odours among flowers arise from interactions with their pollinators.

◇ Can you suggest from Figure 2.13 and Table 2.1 (Section 2.3.1) when the evolution of such reciprocal and elaborate relationships between plants and insects began?

◆ In general, insects do not play specific roles in the reproduction of non-flowering plants, so such interactions did not begin until the angiosperms evolved in the Cretaceous.

By the time they began to play a specific role in plant reproduction, insects had already been living on land for 300 Ma, and almost all the major orders were well-established. Thus it is not surprising that members of many different orders, including Lepidoptera, Hymenoptera, Diptera and Coleoptera should have evolved independently as pollinators of angiosperm plants.

The majority of insects, many nematodes, gastropod molluscs, mammals and birds are herbivores: they destroy the plant's tissues without benefiting it in any way. Certain features of the plants, such as tough, spiny, distasteful or toxic leaves, stems, roots and seeds, play little or no role in the plants' own metabolism but protect them from being eaten. Many insects, particularly caterpillars (larval moths and butterflies), bugs and beetles are only found on one or a few closely related species of plant, and many can feed on only one kind of plant tissue (e.g. the seed or young leaves). Special, sometimes unique, features of the metabolism and life cycle of such herbivores specifically equip them to eat particular plants in quantities that would be toxic to most other animals. Some examples of such adaptations, and how they can be exploited by humans, are described in Chapter 3, Section 3.4. In many cases, the insect's feeding mechanism, digestion and metabolism are so precisely adapted to deal with the mechanical structure and chemistry of particular leaves, roots or seeds that it can thrive only on its own food plant. Such specific, reciprocal adaptations between a herbivore and its particular food plant (or between a predator and its prey) are among the factors that lead to the establishment of a greater number of species, many of which may appear only slightly different but which are adapted to a different diet or habits.

2.5 CONCLUSIONS: WHY ARE THERE SO MANY SPECIES OF ORGANISMS?

This topic is the subject of active research and there is still no complete answer to the question. Nonetheless, some factors that promote the evolution of species diversity have been identified, among them inherent properties of the organism's basic body plan, and ecological relationships such as symbiosis. Soft-bodied organisms such as cnidarians and platyhelminths, in which most of the metabolically active tissues are superficial, are clearly not likely to prosper in dry habitats; insects and tetrapod vertebrates with internal respiratory surfaces are also more likely to be successful on land than crustaceans and fish. However, as pointed out in Section 2.2.4, there are many exceptions to such expectations and generalizations risk becoming after-the-fact rationalizations.

A few of the many ecological factors that promote species diversity have been described in this chapter. Thus a single species of animal or plant, particularly a large one, may harbour many different parasitic species and the parasites may themselves be parasitized by additional species (see Section 2.4.2). In many cases, a particular species acts as a pollinator for or feeds on only one particular tissue of only one species of plant or successfully parasitizes only one host. Each species has features that equip it to prosper only in association with one or a small number of other species. For example, even a small plant such as milkweed (*Asclepias*) is visited by at least twenty different species of arthropods, two of which, the monarch butterfly and its caterpillar (*Danaus*), and the milkweed bug (*Oncopeltus*), feed exclusively on certain tissues of that particular species of plant. Some species of large tropical trees are the basic source of food for up to a thousand species of insects, many of them living only on that particular kind of tree. The total number of species of animals is far greater than that of multicellular plants. Arthropods are by far the most abundant phylum of animals, and the great majority of them are insects. One of the reasons for their huge diversity is that most of the major groups of insects are herbivores, many of them specialized for eating different plant tissues (bark, roots, leaves, pollen, etc.) and many others are parasites, at least for part of their lives.

Summary of Sections 2.4 and 2.5

Symbiosis is prolonged, intimate association between organisms of different species. A symbiotic relationship is described as mutualistic if biologists can demonstrate that both partners benefit from the association; if the relationship is detrimental to one partner, it is called parasitism. Mutual interdependence also occurs between free-living organisms. Specific and elaborate interactions between organisms promote species diversity.

Question 8 (*Objective 2.9*) Which of the following statements are generally *true* about symbiosis?

(a) Mutualism always evolves to become parasitism as one symbiont contributes less to the partnership than the other.

(b) Symbiotic species are biochemically simpler than free-living species in the same major group.

(c) Symbiosis has evolved in only a few groups of organisms.

(d) Ectoparasites are less specialized to their way of life than endoparasites.

(e) Parasitism hastens the extinction of both parasite and host.

(f) Mutualistic symbionts are dependent upon each other for survival.

Question 9 (*Objective 2.10*) Select from the list of features present in some plants and in some or all insects:

(a) Features that adapt the organisms for life on land

(b) Features that are functionally related to insects feeding on plants

(c) Features that are functionally related to pollination of plants by insects

Plant features	*Animal features*
Scented flowers	Tracheae
Brightly coloured flowers	Pupal stage in development
Flowers that produce nectar	Mechanically stiff exoskeleton
Leaves containing toxic substances	Exoskeleton impermeable to water
Wood	Two pairs of wings
Seed formation	Large eyes
Impermeable covering on leaves and stems	Sucking mouthparts
	Biting mouthparts

OBJECTIVES FOR CHAPTER 2

Now that you have completed this chapter, you should be able to:

2.1 Define and use, or recognize, definitions and applications of each of the terms printed in **bold** in the text.

2.2 Explain how the functions of biological structures can be identified correctly. (*Question 1*)

2.3 Recognize and be able to avoid teleological ideas and statements when thinking or writing about biology. (*Question 2*)

2.4 Explain the major features of life cycles in which there is a larval stage and recognize and describe some common types of larvae. (*Question 3*)

2.5 Explain the relevance of larvae to the interpretation of evolutionary relationships. (*Question 4*)

2.6 List some features of the structures and habits of organisms that promote or impede fossilization and explain the kinds of information that can be obtained from the study of fossils. (*Question 5*)

2.7 Describe how comparative and historical information are integrated in attempts to explain the origins and diversifications of major groups. (*Question 6*)

2.8 List the major differences in structure and life cycle of aquatic and terrestrial organisms and outline the major steps in the evolution of terrestrial vertebrates. (*Question 7*)

2.9 Describe the essential features of mutualism and parasitism, and list some adaptations of the structures and life cycles of symbionts. (*Question 8*)

2.10 Explain how species-specific interactions can promote species diversity, using examples from interrelationships between insects and plants. (*Question 9*)

COMPARATIVE AND EXPERIMENTAL BIOLOGY

3.1 COMPARATIVE AND EXPERIMENTAL METHODS IN RESEARCH

Although physical and chemical methods for detecting and quantifying biological processes have been and still are improving very rapidly, many of the features of living systems that biologists would like to know about cannot be measured conveniently or accurately (and often not at all) in the species (e.g. humans and domestic livestock) that people most want to understand. It is not just 'higher functions' and more abstract entities like 'pain' or 'thought' that cannot be quantified; we still do not have suitable techniques for measuring more mundane physiological processes such as the route taken by blood flowing through muscle and skin during strenuous exercise, or the chemical processes involved in tasting sugar or salt, in humans and other large mammals. Much of our knowledge about the basic biological mechanisms comes instead from the study of other kinds of organisms. The bacterium, *E. coli* (Chapter 1, Section 1.2.1), the fruit-fly, *Drosophila* (Chapter 1, Section 1.6.4) and the rat have been most extensively investigated, but important insights that are directly relevant to human physiology have arisen from the study of fungi, maize, earthworms, frogs, deep-sea fishes, lobsters, giraffes, migratory birds and many other species.

There are two principal ways of investigating biological problems: in the **experimental approach**, two groups of organisms (or batches of isolated tissues) as similar as possible in sex, age, genetic make-up and nurture are treated in different ways; one or a few anatomical structures, physiological processes or behaviours are measured accurately and the data from the two groups are compared. Another approach is **comparative**—the study of functionally comparable processes in organisms with contrasting structure and ancestry. Instead of causing differences in specimens that were bred and raised to be similar and then studying the consequences of such intervention, the investigator compares organisms that are naturally different. From the eighteenth to the mid-twentieth centuries, comparative biology was enormously successful; the many advances in both information and concepts, such as theories about the mechanisms of evolution, made possible a huge range of economically important discoveries, among them rubber and antibiotics. However, in the late twentieth century, comparative biology is somewhat out of fashion. This chapter is about how a knowledge of the diversity of organisms can guide and complement laboratory and experimental studies. In Section 3.2, we describe some applications of the comparative method to basic biological research. The example chosen is the elucidation of the basic mechanism of transmission of signals within and between nerves. The story so far has involved more than a dozen different species of wild animals and plants and illustrates the essential role played by a knowledge of the diversity of organisms. Biologists who are familiar with a wide range of organisms can exploit their special properties that may turn out to be exceptionally amenable to study with available techniques.

Section 3.3 is about the uses and limitations of laboratory animals (e.g. rats and mice) for studying normal and abnormal physiological processes and the extrapolation of experimental results from laboratory species to rare, dangerous or very large species, and humans, pets and farm animals, which, for practical and ethical reasons, are very difficult to study experimentally. We also explain how information from wild animals guides and complements experimental studies. Understanding the biochemical mechanisms and the biological relations between organisms is of more than academic interest: as well as almost all foods, many of our most important building materials, textiles and medicines are derived from wild organisms. Section 3.4 is about the functional and physiological reasons for the presence of these economically important components, and their identification and exploitation for human use.

3.2 THE STUDY OF THE NERVOUS SYSTEM

Until the early nineteenth century, biologists gave the name neurons to all the structures that we now call nerves and tendons. All the information available to them came from observations on dead, often pickled specimens, and provided no basis for distinguishing between different kinds of white, fibrous structures associated with the muscles and skeleton. The microscopic appearance of nerves and tendons are also quite similar: both tissues consist of bundles of round structures, which in nerves are projections of a special kind of cell to which the term **neuron** is now exclusively applied, and in tendons are extracellular fibres composed almost entirely of one kind of protein (collagen). As in other cells, the cytoplasm of neurons contains numerous different enzymes and structural proteins and are bounded by a highly organized cell membrane.

The eighteenth-century Italian anatomist, Luigi Galvani*, was the first scientist to suggest that electricity might play a central role in animal movement: he noticed that the legs of freshly killed frogs would twitch or kick if an electric current from a condenser was discharged between them. The effect diminished rapidly and disappeared completely within a few hours after death. Further observations on living animals and freshly excised tissues pinpointed nerves and muscles as the structures most involved in generating 'bioelectricity' and in responding to external sources of current.

3.2.1 Nerve impulses

By the 1920s, most biologists agreed that nerves communicated with each other, and probably also with muscles, by means of 'electrical impulses'. Propagation of these impulses could be prevented permanently by cutting or pinching the nerves, or temporarily by cooling them. In many cases, nerve impulses could be initiated artificially by a variety of mechanical and electrical means, of which the most effective was applying brief shocks of alternating current. Several eminent physiologists successfully used this procedure to study connections between neurons and the neural pathways that underlie some simple actions in frogs, rats and cats.

*Although primarily a professor of medical and comparative anatomy, Galvani also made significant contributions to physics, and gave his name to the galvanometer, a device for measuring small electric currents.

Neurons, particularly those of vertebrates, are small, delicate cells usually consisting of a cell body that contains the nucleus, and one or more long processes called axons. Almost all peripheral nerves in vertebrates contain from dozens to thousands of long parallel axons, often called nerve fibres. Figure 3.1 is a cross-section of a large peripheral nerve, the sciatic nerve, in the hindleg of a mouse, magnified about 11 000 times. The axons vary greatly in thickness, the majority being 1 μm or less in diameter and the largest only about 4 μm thick. Most of the small ones are tightly packed in bundles and the larger ones are surrounded by a supporting tissue, myelin. Such small axons cannot easily be separated from each other and from the supporting tissue without damage and so cannot be studied individually. Therefore, although the properties of bundles of nerve fibres had been extensively studied, as recently as 1950 very little was known about how nerve impulses were generated and propagated in a single neuron. The way to solve this problem may seem obvious with hindsight, but was a major breakthrough at the time: don't bother with small, fragile, experimentally inconvenient nerves, study neurons that are large, robust and accessible.

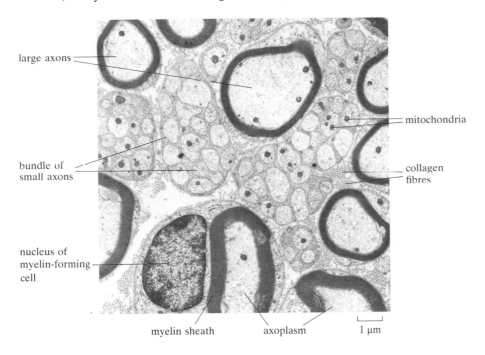

large axons

mitochondria

bundle of small axons

collagen fibres

nucleus of myelin-forming cell

myelin sheath   axoplasm   1 μm

Figure 3.1 Cross section of the sciatic nerve in the hindleg of a mouse, as seen in the electron microscope, magnified 11 600 times. The majority of axons are small and are grouped together in bundles. The larger axons are enclosed in a fatty sheath called myelin, which is formed by special non-conducting cells. The nucleus of one such cell is visible, and indicates how small the axons are.

Many invertebrate neurons, particularly those in the brain and sense organs, are also small and delicate and hence are as technically inconvenient to work with as those of vertebrates. However, in the early 1930s, a zoologist, J. Z. Young, drew attention to some exceptionally large neurons in the nervous system of the squid *Loligo* and some other fast-swimming cephalopods such as cuttlefish. Figure 3.2 shows the arrangement of the principal 'giant' neurons; some run from the eye to the brain, but the majority, including the largest, which are about 1 mm in diameter and up to 100 mm long, originate from a mass of nervous tissue on the inner wall of the mantle called the stellate ganglion and innervate the musculature of the mantle (see Chapter 1, Section 1.6.5).

The axons of the giant neurons of the squid are large enough and robust enough for fine electrodes to be inserted through the cell membrane and into the cytoplasm. When connected to suitable apparatus, such electrodes can be used to measure electrical processes, including current flow and the voltage

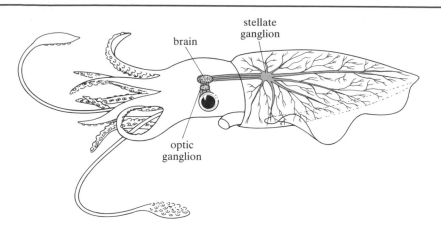

Figure 3.2 The arrangement of the major giant neurons (shown in red) in a squid. Note that the axons in the mantle are tapered, and that near the stellate ganglion, the longer ones are thicker.

between the inside and the outside of the cell. Like all molluscs, cephalopods are poikilothermic; the species of squid and cuttlefish used are native to middle-depth waters of temperate seas, so their physiological processes function best at about 10–15 °C. Furthermore, the molluscan blood system consists of a haemocoel (see Chapter 1, Section 1.6.5) in which many of the tissues are surrounded by blood at low pressure. Such conditions are readily imitated in the laboratory: the tissues are bathed in a cooled solution of salts and nutrients through which oxygen (or just air) is bubbled.

By the early 1950s, physiologists had perfected methods for dissecting out single giant axons and studying them in isolation. The nerve impulse was shown to be a transient flow of ions (atoms or molecules that contain more or fewer electrons than protons, and so carry an electric charge) through the axon membrane, producing a brief change in voltage across it, called the **action potential** (Figure 3.3). Under suitable conditions, an isolated giant axon would remain healthy for hours and generate many thousands of action potentials, more than enough for physiologists to make the detailed measurements that they needed to unravel the complicated mechanisms of their formation and propagation. Small positively charged ions, particularly those of sodium and potassium, were of special interest because they were thought to carry the currents that flowed along and through the neuron membrane. The giant axons turned out to be so tough that the intracellular contents could be squeezed out like a tube of toothpaste, leaving just the cell membrane. The 'empty' axons could be refilled with artificial solutions of precisely known

Figure 3.3 A single action potential as displayed on an oscilloscope, an instrument that acts as a fast-responding voltmeter. The display shows the changes in voltage across a neuron membrane (vertical axis) in time (horizontal axis) taking place at a single point on it. Note that by 9 ms after the start of the action potential, the voltage across the membrane has returned to its original value.

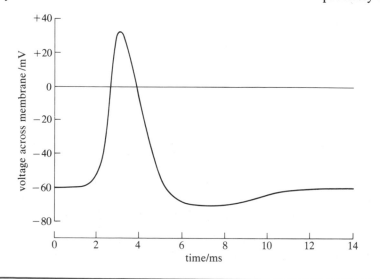

chemical composition, enabling physiologists to study, with unprecedented precision, the movements of small molecules across the axon membrane and mechanisms that control them.

Calculation of the net change in concentration of sodium and potassium ions after isolated axons had been stimulated artificially to produce thousands of action potentials, suggested that about 10^{-12} moles of each ion enter or leave the neuron through each square centimetre of active membrane during a single action potential, probably through special channels. In the 1950s, it was virtually impossible to measure such small quantities of sodium and potassium separately using chemical methods alone. The prospects did not look good for following the movements of ions during a single action potential or elucidating further details of the molecular mechanisms that control the abrupt, transient changes in the neuron membrane that underlie its formation. However, there were electronic instruments sensitive enough to measure small changes in electric charge that would be caused by the movement of such small numbers of positive ions. So, if movement of one kind of ions could be prevented totally, movements of the other could be followed as electrical changes. This is where the Japanese puffer fish *Tetraodon* and a small South American tree frog *Phyllobates terribilis* entered the story.

The genus *Tetraodon* (Figure 3.4) includes several species of slow-swimming fishes that live in coastal waters and around reefs, feeding on coral, shelled molluscs and other slow-moving invertebrates. Their common name, puffer fish, derives from their habit of inflating themselves when alarmed by swallowing water, and, in the case of shallow-water species, air; predators cannot easily swallow the stiff, distended body and do not molest them. Their flesh is highly prized as a delicacy (called fugu) in Japan and Southeast Asia. However, although the blood and muscles are appetizing and nutritious, the gonads, liver and intestines of many puffer fish are lethally poisonous to humans (and all other vertebrates on which they have been tested), causing death by inactivation of the neurons that control breathing movements. (In Japan, only cooks who have been specially trained to identify and remove the toxic tissues are allowed to prepare fugu.) The toxic ingredient was isolated in 1950 and named tetrodotoxin. Biologists tried adding tetrodotoxin to the solution around an isolated giant axon while stimulating it repetitively for several minutes, and found that the movements of sodium ions were blocked, while those of potassium ions were unaltered. They concluded that sodium and potassium ions move through different channels in the membrane. This unexpected feature of neuron membranes turned out to be fundamental to the elucidation of the mechanism of action potential formation. The effects on action potential formation of several other natural and synthetic toxins were investigated, among them toxins produced by *Phyllobates terribilis*.

Phyllobates and the related genus *Dendrobates* are small, brilliantly coloured frogs that do not jump or run away if approached or handled. Predators (such as snakes and birds) do not molest them because glands on their backs secrete a foam that is dangerously toxic if swallowed. Scientists collected some of the foam and studied its effects on isolated tissues, including neurons. They separated its components (using chemical isolation and purification techniques) and showed that the active ingredient was a nerve poison that they called batrachotoxin. Studies of its effect on isolated neurons showed that batrachotoxin also interfered with the movement of sodium ions across the neuron membrane during action potentials, but in a different way from tetrodotoxin. Further studies revealed that tetrodotoxin specifically blocks the channels through which the sodium ions pass into the neuron at the beginning of the action potential, while batrachotoxin prevents the termination of movements of sodium ions at the end of the action potential.

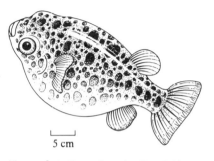

5 cm

Figure 3.4 Tetraodon (puffer fish), a highly poisonous teleost fish (order Tetraodontiformes).

◇ Would it have been possible to use synthetic drugs that block particular components of the action potential as specifically as the natural toxins?

◆ Almost certainly not. It would be impossible to synthesize a substance that reacted specifically with channels until they had been identified and isolated. The two kinds of channels are distinguished mainly by the differences in their interaction with the natural toxins.

Although it has only recently come to the attention of biologists, the Chocó Indians in the Amazon rainforest have kept *Phyllobates* frogs in captivity for thousands of years to use their secretions for coating the tips of their arrows.

◇ Why would the skin secretion of *Phyllobates* frog be effective for poisoned arrows?

◆ It contains batrachotoxin which interferes with the formation of action potentials, and so prevents proper transmission of signals from the brain to the muscles, paralysing the prey within seconds.

In fact, batrachotoxin is so deadly that one small frog yields sufficient poison for about fifty arrows. Many plants also contain substances that interfere with chemical and electrical processes in the nervous system and some were put to practical use long before their active ingredient was isolated or the mechanism of their action understood. For example, extracts of the fruits and bark of a large tree, *Strychnos*, was widely used for hunting wild mammals and birds in tropical Asia; modern chemical and physiological techniques revealed that they contain a powerful inhibitor of neural activity called strychnine, which acts primarily on the junctions between neurons (see Section 3.2.3).

The ionic currents produced by the presence of an action potential in one area of membrane stimulate another action potential in an adjacent patch of membrane, so once started, the signal propagates along the neuron. Once they knew more about the biochemical processes involved in the formation of action potentials, and had designed better apparatus with which to study them, physiologists examined smaller neurons, including those of mammals. They were pleased to find that the basic mechanisms of formation and propagation of action potentials were remarkably similar in a wide range of animals, including humans and other mammals. (Although the details of how action potentials propagate around the complicated shapes of elaborately branched neurons are rather different in molluscs and mammals.) One of the most variable features was the speed of propagation; the conduction velocity of action potentials in the squid giant axons turns out to be among the highest of all invertebrate neurons.

◇ What role does this unusual neuron play in the physiology of the squid? Can you make any suggestions, referring to Figure 3.2 and Chapter 1, Section 1.6.5?

◆ The giant neurons connect the stellate ganglion (a concentration of nerve tissue) with the muscles of the mantle. Squids swim by 'jet propulsion', which is most effective when all the swimming muscles contract exactly simultaneously. Action potentials travel faster along large neurons than along smaller ones of similar anatomical structure, so the most likely explanation for the presence of the giant axons is that they coordinate rapid, synchronous activation of the swimming muscles.

The 'jet propulsion' escape behaviour of the squid begins when information passes from the eyes via the optic ganglia to the brain (Figure 3.2). Large neurons convey signals from the brain to the stellate ganglion in the mantle, where they excite the long giant axons that innervate the muscles of the mantle. As indicated in Figure 3.2, the longest axons are also the thickest, and hence propagate impulses at the fastest speed. The conduction velocity of nerve impulses and the anatomical distribution of the axons are arranged so that signals that start in the stellate ganglion travel faster to more distant muscles than to adjacent ones, reaching all of them at the same time. This arrangement ensures that all the muscles contract simultaneously, producing a powerful jet of water that enables the squid to move very fast. Biologists have simply exploited the technical advantages of these exceptional neurons to develop concepts and methods for studying an important and basic physiological mechanism.

3.2.2 Transmission between nerves and nerve-like tissues

Although electrical signals can be detected throughout the nervous system, indicating that action potentials must somehow be transmitted from neuron to neuron, measurements made on the squid giant neurons showed that action potentials travelled along the entire axon without attenuation, but did not necessarily propagate to adjacent neurons. The study of sections of nervous tissue revealed localized junctions between neurons, called **synapses**, that had a number of intriguing features, including thickened cell membranes on both sides of the junction and small vesicles on one side only (Figure 3.5). Synapses were suspected of playing a central part in the transmission of action potentials between neurons and physiologists were keen to find out more about how they worked.

Unfortunately studying intact, living synapses presented serious practical problems. The basic mechanisms of synaptic transmission of signals in the nervous system would probably be best studied in a single large, easily localized junction between only two neurons. However, as you can see from Figure 3.5, the junctions between neurons, including those on the giant axons, are very small; furthermore, they tend to occur in dense clumps, surrounded by other neuronal tissue, making it almost impossible to determine what is connected to what. Most large neurons in living, intact nervous systems are highly branched and often form connections with scores, sometimes thousands, of other neurons. To make matters even worse, anatomical

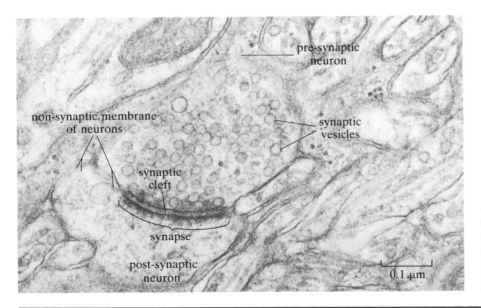

Figure 3.5 Electron micrograph of a synapse between neurons in the brain of a newly hatched chick. The Figure shows parts of two neurons. The upper one is pre-synaptic; the lower one is post-synaptic. Only part of the neuronal membrane forms a synapse.

studies suggested that the structure, arrangement and probably therefore the physiological properties of synapses between neurons are quite variable between specimens, and may change with the age and recent experience of the animal.

However, the external membrane around another tissue, muscle, is similar in several respects to neuronal membranes. Most muscles are made up of from a handful to many thousands of **muscle fibres**, each of which consist of cytoplasm containing the contractile proteins and several nuclei, surrounded by an external membrane. Using electrodes similar to those developed for studying the squid giant axon, physiologists found that many muscles, including most limb muscles of vertebrates, generate propagating action potentials similar to those of the squid giant neuron. Furthermore, the neurons that convey signals from the central nervous system to the muscles, called motor neurons, seemed to be among the largest and least variable of the whole nervous system. Most of the thicker neurons in Figure 3.1 are probably axons of motor neurons that carry signals from the spinal cord to the muscles of the hindleg. Better still, the anatomists were able to tell the physiologists that motor neurons made contact with the muscles at a small number of relatively large, synapse-like connections called **neuromuscular junctions**. All in all, neuromuscular junctions seemed to be promising structures on which to begin investigations into how neural signals travel from cell to cell and, during the 1950s and 1960s, they were studied intensively.

Vertebrate neuromuscular junctions

Figure 3.6a is an electron micrograph of a vertebrate neuromuscular junction, the major features of which are shown diagrammatically in Figure 3.6b.

◇ Compare the size and shape of the neuromuscular junction (Figure 3.6a) with the size and shape of the neuron–neuron synapse (Figure 3.5).

◆ The neuromuscular junction (Figure 3.6a) is much larger than the synaptic area of the neuron–neuron synapse (Figure 3.5). The motor neuron terminal contains more mitochondria and vesicles than the presynaptic nerve terminal, and the membrane on the muscle side of the neuromuscular junction is invaginated.

In frogs (and in birds and mammals), there is normally only one neuromuscular junction on each muscle fibre. The action potential travels down the motor neuron to the neuromuscular junction, where it starts a similar action potential that propagates over the membrane of the muscle fibre and initiates contraction. The hindlegs of frogs are relatively very long and contain several powerful muscles that are involved in swimming and leaping. Because the legs are long and muscular, the nerve containing the motor neurons running from the spinal cord to the muscles of the lower leg is also relatively long and stout, and it is relatively easy to remove the whole nerve and the muscle intact from a freshly killed frog.

Although much larger than neuron–neuron synapses, at only about 1 μm in diameter the neuromuscular junction is still very small compared to the dimensions of the physiologists' equipment! Nonetheless, biologists developed methods for locating the tiny neuromuscular junctions on the muscles of a freshly killed frog and inserting fine electrodes into the motor neuron terminal and the muscle side of the neuromuscular junction at the same time. They were thus able to apply controlled stimuli to the motor neuron and to record electrical events in both sides of the neuromuscular junction. They found that sodium and potassium ions were important carriers of currents flowing across membranes, but that the mechanisms that controlled ion movements at the neuromuscular junction are very different from those operating in the non-synaptic regions of neuron and muscle fibre membranes. In particular, a chemical **transmitter** called **acetylcholine** was found to play a central role in generating the action potential on the muscle membrane. Acetylcholine is synthesized by the motor neuron, stored in the little vesicles visible in the terminal in Figure 3.6, and released into the space between the neuron and the muscle. A special area of muscle membrane under the neuromuscular junction generates signals when acetylcholine is bound to it. In the neuromuscular junction, acetylcholine normally remains active for only

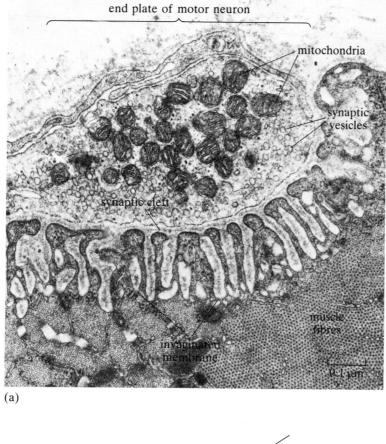

(a)

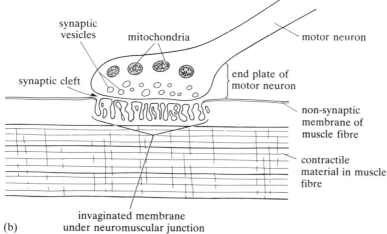

(b)

Figure 3.6 (a) Electron micrograph of a neuromuscular junction between a motor neuron and a skeletal muscle of an adult mouse. (b) Diagram of the main features of a neuromuscular junction.

a few milliseconds (1 ms ≡ 10^{-3} s) before an enzyme, **acetylcholinesterase**, which is present in the gap between the neuron and the muscle, splits the molecule into inactive components, some of which are then reabsorbed by the motor neuron terminal.

It was obviously important to know more about the mechanisms by which acetylcholine binds to the membrane on the muscle side of the neuromuscular junction and causes it to generate electrical signals. One of the most useful tools in this investigation was the venom of kraits of the genus *Bungarus* (Figure 3.7), particularly the southeast Asian species *B. multicinctus*. Kraits are snakes of the cobra family (Elapidae) that feed almost exclusively on other snakes, including large, powerful species such as king cobras. Careful observation of the process of predation revealed that kraits seize their prey and hold it firmly until it dies (in contrast to many other poisonous snakes, such as vipers and rattlesnakes, which bite the prey, let go, and follow it until it collapses). Prey bitten by kraits stop struggling within seconds and even animals as large as humans die within half an hour of being bitten.

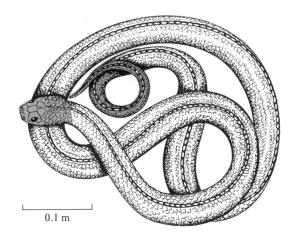

Figure 3.7 The red-headed krait, *Bungarus flaviceps*, native to India.

0.1 m

◇ What can you conclude from these observations about which tissues are most likely to be affected by the venom?

◆ Motor neurons and/or muscles. Damage to the lungs or the blood system would take longer to cause paralysis.

The principal ingredients of krait venom were isolated and their effects on vertebrate nerves and muscles were studied. One major component, named bungarotoxin, was found to block transmission of signals across the frog neuromuscular junction. Further studies, involving several ingenious procedures, among them feeding captive *Bungarus* on radioactive prey so that it would synthesize radiolabelled venom, showed that bungarotoxin blocks transmission because acetylcholine binds to the muscle fibre membrane only at certain points, called **receptor sites**, to which bungarotoxin also binds readily. Receptor sites that are 'occupied' by the foreign substance cannot bind acetylcholine, so the membrane under the neuromuscular junction does not respond to the transmitter, even when it is present at very high concentrations, and transmission fails. Neither tetrodotoxin nor batrachotoxin (Section 3.2.1) have any detectable effect on neuromuscular transmission, even when they are administered in large doses, and bungarotoxin has no effect on membranes other than those under neuromuscular junctions.

◇ What can you conclude from this information about the mechanisms that underlie action potentials and neuromuscular transmission?

◆ Propagation of action potentials within neurons is fundamentally different from transmission of signals across neuromuscular junctions.

In fact, the two processes involve chemically different channels as well as taking place on structurally different areas of the cell membrane.

These natural toxins played a unique and essential role in demonstrating that different structures and mechanisms underlie the ability of membranes to react to acetylcholine or to support an action potential.

◇ Why would a substance that binds irreversibly to acetylcholine receptors quickly paralyse and kill vertebrates?

◆ The poison would immobilize all the skeletal muscles, including the respiratory muscles, by blocking transmission of signals from the motor neurons. The prey would be unable to struggle and would quickly suffocate. Clearly, from the snake's point of view, such a substance is an effective means of killing large, potentially dangerous prey.

When sequestered in the venom glands (which probably evolved from modified salivary glands) above the fangs (modified teeth), the toxins are harmless to the animal that produces them. However, most snakes can be poisoned by their own venom if it enters their blood, either through damage or disease to the fangs or poison glands, or by the snake inadvertently biting itself.

Several other natural and synthetic substances inhibit various steps in neuromuscular transmission. Many, like bungarotoxin, cause irreversible changes in the tissues but others are inactivated by normal metabolic processes, or can be removed by antidotes. For example, curare, which is found in several species of plants including the South American tree *Chondodendron tomentosum*, is widely used as a muscle relaxant for surgery on humans and other vertebrates. Like bungarotoxin, it binds to acetylcholine receptor sites but not with such high affinity as the snake venom; some of it remains in the blood circulation and is broken down in the liver to harmless products, which are excreted by the kidney. Reversal of the effects of curare can be accelerated by administering neostigmine, which promotes the release of acetylcholine at neuromuscular junctions.

Thus the study of the frog neuromuscular junction revealed how an action potential in the neuron generates an action potential in another neuron-like cell, the vertebrate muscle fibre. Electron microscopy revealed many similarities between neuromuscular junctions and neuronal synapses: there are vesicles in presynaptic terminals of the neurons (Figure 3.5) similar to those in the motor neuron terminal of the neuromuscular junction (Figure 3.6) and the membranes on both pre-synaptic and post-synaptic sides of the junction are thickened. When electrical recording techniques became sufficiently refined, many physiological similarities between neuromuscular junctions and synapses were also demonstrated. Thus detailed study of amphibian neuromuscular junctions provided essential background to the study of mammalian neuronal synapses.

However, because it consisted of a single motor neuron connected to a single muscle fibre, study of the frog neuromuscular junction could reveal nothing about how inputs from several synapses might be integrated together. Furthermore, in frogs (and all other vertebrates), oxygen and nutrients reach the muscles and nerves via the blood which is pumped under pressure through the fine blood vessels that permeate the tissues (see Chapter 1, Section 1.6.7). Unless perfused artificially (which requires elaborate apparatus and is often technically difficult), vertebrate tissues quickly become depleted of oxygen and other essential metabolites as soon as their natural blood supply is impaired, or when they are removed from the body.

Arthropod neuromuscular junctions

Many large crustaceans (class Malacostraca, order Decapoda), including crayfishes, lobsters and most crabs (Plate 11g), seize prey and tear up food using appendages that terminate in powerful claws, called chelae (Chapter 1, Section 1.6.4 and the Appendix). Some species also use the same appendages in courtship and territorial defence displays. In many lobsters and crabs, the chelae are asymmetrical (Figure 3.8), with one chela being long and narrow with a sharp, saw-like edge, and the other being short and stout and lined with hard, blunt nodules suitable for crushing. The muscles that operate the opposing segment of the claw are capable of finely graded, precisely controlled movement; if you have ever been attacked by such an animal, you will know that the claws can inflict a powerful and prolonged pinch. If you have ever eaten lobsters or crabs, you will know that the chela muscles are large and meaty, and are easily exposed by lifting off the stiff cuticle that surrounds them.

◇ What features of the arthropod respiratory and circulatory systems are likely to make them more robust during prolonged experiments than vertebrates?

◆ In vertebrates, there is only one 'central' respiratory surface (lungs or gills) and the heart pumps the blood to tissues in closed vessels and the tissues are quickly deprived of oxygen if the heart and lungs (or gills) are removed or impaired, or the blood pressure falls. In contrast, arthropod muscle and nervous tissues obtain nutrients from the blood in the haemocoel. Oxygen is obtained from adjacent tracheae in terrestrial arthropods and from thoracic or abdominal gills in aquatic groups including crustaceans (Chapter 1, Section 1.6.4; Chapter 2, Section 2.3.2). Although large arthropods have a heart (Figure 1.24), the blood pressure is always much lower than that of vertebrates.

It was these features that attracted physiologists: the chelae of decapod crustaceans contain large, surgically accessible muscles that continue to function almost normally when isolated from a freshly killed lobster or crayfish and bathed in artificial fluids.

Both anatomical and physiological studies of these muscles made it clear that the arrangement and properties of the motor neurons that controlled them so effectively are quite different from those of frogs. Two or three, sometimes more, motor neurons are connected to each muscle fibre, and each forms scores of neuromuscular junctions distributed over the whole surface of the muscle fibre (and sometimes branches of a single motor neuron innervate several muscle fibres). Before they branch out near the muscle, the axons of these motor neurons are large enough to be teased out and each stimulated separately. Unlike the membrane of vertebrate muscles, the external membrane of most arthropod muscles is incapable of generating propagated action potentials.

It turned out that applying the same stimuli to different combinations of motor neurons produced different movements. Some produced brief, fast movements, others slow, prolonged activity; stimulation of a third type of

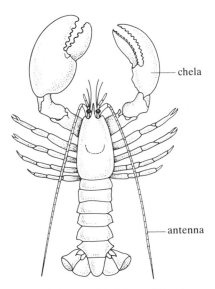

Figure 3.8 An adult lobster. The short, powerful left chela is adapted for crushing food (probably other arthropods and shelled molluscs). The more slender right chela is adapted for tearing.

motor neuron curtailed, delayed or even prevented contraction that had been initiated by stimulation of other motor neurons on the same muscle. This third type is called an **inhibitory neuron**. The different kinds of motor neuron also generated contrasting electrical signals on the muscle membrane and involved different chemical transmitters. For example, acetylcholine is not found in these neuromuscular junctions, but one of the major inhibitory transmitters is γ-aminobutyric acid (usually abbreviated to GABA). Thus the muscles of the crustacean chela offered an ideal model of neuronal synapses that integrate signals from several different neurons.

The crustacean neuromuscular junction provided the first reasonably complete picture of how signals arriving from several different sources at more or less the same time are combined to generate signals in the postsynaptic cell. Techniques and concepts developed from such research were applied to the study of neuronal synapses, including those in the mammalian nervous system. Vertebrate neuronal synapses turned out to be similar in many ways to arthropod neuromuscular junctions, and many of the chemical transmitters found in vertebrates, among them acetylcholine, proved to be active in the central nervous system of other invertebrates, particularly insects. As laboratory animals, insects have many advantages comparable to those of crustaceans: insect tracheae continue to function when isolated from the rest of the respiratory system, most species do not have blood-borne respiratory pigments (such as haemoglobin) and the blood system does not normally function under pressure. In general, isolated insect tissues survive much longer than those of vertebrates, making it much easier to study them experimentally. During the last thirty years, insects, particularly large species such as locusts and cockroaches, have been widely used as laboratory animals in which to investigate basic neurobiological mechanisms.

3.2.3 Exploiting the differences

◇ Would lobsters and crayfish be paralysed by a bite from *Bungarus*?

◆ No. Bungarotoxin binds specifically to receptor sites for acetylcholine, which are not present at crustacean neuromuscular junctions, because acetylcholine does not act as a transmitter. Therefore, bungarotoxin would not affect transmission at these neuromuscular junctions and the arthropods would not be paralysed.

A large crab could injure a snake, whose mechanically weak jaws cannot subdue struggling prey. *Bungarus* (like most other elapid snakes) does not tackle invertebrate prey; in fact, kraits rarely strike at anything other than their natural food, and so are notoriously difficult to keep in captivity.

Biologists exploit such intrinsic differences in the effects of toxins on different kinds of organisms in order to develop techniques or materials that destroy or nurture particular species. Take the example of insecticides, toxic chemicals that are sprayed over plants, humans and livestock in order to eliminate the insects and other arthropods (e.g. mites) associated with them. Although acetylcholine is rarely found at arthropod neuromuscular junctions, it is important as a transmitter at synapses between neurons in the central nervous system, particularly that of insects. One of the first really effective insecticides was DDT (dichlorodiphenyltrichloroethane), a synthetic substance that interferes with the normal functioning of acetylcholinesterase, the enzyme that breaks down acetylcholine.

◇ How would you expect an insect poisoned with DDT to behave?

◆ Without acetylcholinesterase, acetylcholine released at synapses is not broken down, and so continues to promote the formation of action potentials in neurons. High frequencies of action potentials in the motor neurons causes continuous, simultaneous contraction of all the muscles, making the legs and wings tremble.

The insects die from exhaustion in a few minutes.

◇ Would you expect DDT to be toxic to vertebrates?

◆ Yes, if their acetylcholinesterase is also inhibited by DDT.

Unfortunately, vertebrate and insect acetylcholinesterase are almost identical, and both are affected by DDT. Being almost insoluble in water but soluble in lipids (i.e. fats and oils), DDT penetrates the fatty external layers of arthropod cuticle relatively easily, and so is absorbed more rapidly by arthropods than by vertebrates. Nonetheless, with prolonged exposure, vertebrates can absorb toxic quantities of DDT.

◇ In which tissues or secretions would you expect DDT and other lipid-soluble insecticides to accumulate?

◆ In fatty tissues and secretions, such as adipose tissue and milk.

Unfortunately, people consume both milk (and milk products such as cream and butter) and adipose tissue (as a major component of meat) in large quantities, and may thereby ingest toxic doses of concentrated insecticide. Carnivorous mammals and birds of prey such as owls, hawks and eagles incur similar risks from eating poisoned insects and other wild animals that have eaten the pesticide. Some sort of insecticide is essential to control harmful ectoparasites such as lice (see Chapter 2, Section 2.4.2) and to protect crops from insect damage, so biologists searched for substances that were toxic to arthropods but harmless or inaccessible to vertebrates. Pyrethrum occurs in the flowers and leaves of certain African daisies. In arthropods, it produces almost exactly the same symptoms as DDT, but more thorough investigation showed that the mechanism of its action is different. Instead of interfering with acetylcholinesterase, it blocks one of the ion channels in the membrane that carries the currents initiated by the binding of the transmitter to the membrane. Furthermore, only arthropod ion channels are blocked by pyrethrum; those of vertebrates are unaffected, probably because the molecular structure of the ion channels in their membranes is significantly different from those of arthropods. Pyrethrum is almost as toxic to arthropods as DDT, but, because it is almost completely harmless to most vertebrates, including fish, wild mammals and birds, humans and domestic livestock, it is much safer for use on farms or in the home.

In this example, the identification of small, and possibly functionally unimportant, physiological differences leads to the development of methods for selectively eliminating one group of organisms without causing harm to others.

◇ Referring to Chapter 2, Section 2.4.3, can you suggest what the natural role of pyrethrum might be?

◆ Many plants produce toxins that deter insects and other herbivorous animals from eating their tissues. Pyrethrum may protect the plant from insects in this way.

Plant tissues, particularly the leaves and roots, often contain many substances that have insecticidal properties, but many are difficult to extract and purify, or have harmful side-effects, or are effective against only a few species of insects.

◇ Referring back to Chapters 1 and 2 and thinking of which other kinds of animals feed mainly or entirely on insects, and how such predation takes place, can you suggest some other biological sources of insecticides?

◆ Spiders and scorpions inject venom into their prey, which are mainly other arthropods, particularly flying insects. The venom paralyses the prey within seconds and so is clearly an effective insecticide.

Furthermore, with a few widely publicized exceptions (e.g. black widow and bird-eating spiders), most spider venoms are harmless to vertebrates. Spider venoms have been extensively studied, both in order to understand more about the natural interactions between the predators and their prey, and because of their relevance to developing safe, effective insecticides. Such research has revealed that many spider venoms interfere with the formation and propagation of action potentials or with synaptic transmission, and that a wide variety of molecular mechanisms are involved.

Summary of Section 3.2

The nervous systems of certain molluscs and arthropods include exceptionally large, physiologically robust neurons that can be isolated and that are more suitable for experimental study than those of vertebrates. The arrangement and properties of motor neurons and muscles in arthropods have important features in common with those of neuronal synapses, but are much more amenable to physiological study. Therefore these invertebrate tissues provide an excellent opportunity for studying the main features of physiological processes and developing concepts and experimental techniques that can then be applied to organisms, such as mammals, that are technically less convenient to study. Certain natural toxins interfere specifically with certain stages of action potential formation and synaptic transmission. Their natural function is to paralyse prey but physiologists use them to study basic neurobiological mechanisms and may exploit their toxicity as insecticides.

Question 1 (*Objective 3.2*) Explain in a few sentences what major discoveries in basic neurobiology have involved the organisms (a)–(c) and why they proved particularly suitable for that investigation.

(a) Frogs   (b) Lobsters and crayfish   (c) Kraits

Question 2 (*Objective 3.3*) Which of the features (a)–(h) make cephalopods more suitable than mammals for studying the basic properties of neurons?

(a) Cephalopods are easier to keep in captivity.

(b) Cephalopods are capable of a wider range of activities.

(c) The cephalopod nervous system is larger.

(d) Cephalopod neurons survive well when isolated and perfused with an artificial bathing solution.

(e) A few neurons in cephalopods are exceptionally large.

(f) All cephalopod neurons are exceptionally large.

(g) Cephalopods have a haemocoel.

(h) Cephalopods are poikilothermic.

3.3 ANIMAL MODELS

Although many biological processes can be elucidated successfully from studies of isolated molecules, cells or organs, some phenomena (particularly those involving interactions between several different organs or different kinds of cells) must be studied *in vivo*. Biological research is often (but certainly not always) directed towards understanding normal physiological processes and the mechanisms of disease in humans and domestic livestock. When studies on these large species are impractical, small animals are often used as 'models'. Because of their importance as animal models, rodents, particularly members of the family Muridae (rats, mice, hamsters and gerbils), are by far the most thoroughly studied small mammals. Most of the 1 700 or so species of wild rodents gnaw on seeds and other dried plant food, but many also occasionally eat carrion and large invertebrates, particularly when breeding. Many species, including rats and the house mouse, readily colonize new environments and exploit new sources of food and they breed very rapidly when food is plentiful, sometimes becoming very abundant. It is these natural features that make murid rodents particularly suitable as laboratory animals: they are small, sociable animals that flourish even at high density. They breed prolifically under simple conditions, reaching sexual maturity a few months after birth, and thrive on a variety of different diets. Rodents have been used successfully for the eludication of many basic physiological processes and for many important applied research projects, but there are limitations and pitfalls to interpreting data obtained from such studies and applying them to other species. Some of these problems may be anticipated from some of the information and concepts developed in Chapters 1 and 2. The examples described in this section are chosen because they illustrate some of the problems that can arise from using laboratory animals as models for human conditions, and how studying a wider range of species can help to overcome them.

3.3.1 Experimental animals in the study of obesity

Obesity is excess adipose tissue; it is not lethal in itself, but it predisposes people (and their livestock) to greater risk of a wide range of disorders, from breaking limbs to heart disease. However, in spite of a huge range of diets and exercise programmes, there is no reliable, lasting cure for this prevalent and distressing disorder.

There has been much research into the properties of adipose tissue and ways to prevent or reverse its growth. A fundamental issue was whether adipose tissue enlarged by proliferation of additional cells, or by expansion of a constant population of cells, or both. Fortunately, such a question seems to be easy to answer in the case of adipose tissue, because at least 90% of its volume is occupied by a single type of relatively large, spherical cells called **adipocytes**. Mature adipocytes are easily recognizable because they are full of fat, and their diameter can be measured easily with an ordinary microscope and from such data the mean volume is easily calculated. If the gross mass of the depot and density of the tissue are known, the numbers of adipocytes in it can also be calculated. Figure 3.9 shows the relationship between adipocyte volume and fatness, the gross mass of adipose tissue expressed as a percentage of the total body mass (to take account of individual differences in lean body mass), in 135 guinea-pigs bred and raised under controlled conditions and fed on a standard laboratory diet. The data are scattered, with the mean volume of adipocytes of some specimens being almost twice as large as that of others of similar fatness. Nonetheless, the data show some general trends that are indicated by the regression lines fitted to the points.

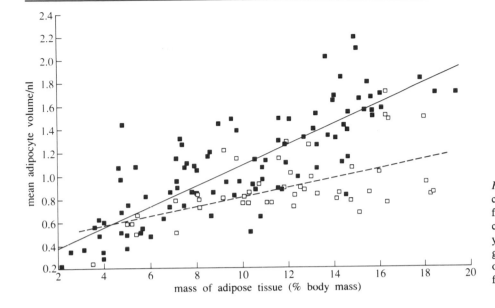

Figure 3.9 Relationship between adipocyte volume ($1 \text{ nl} = 1 \times 10^{-9}$ litres) and fatness (mass of adipose tissue as a percentage of the total body mass) in 97 young (4–10 months old) adult laboratory guinea-pigs (black squares) and in 48 older (over 14 months old) guinea-pigs from the same colony (white squares).

◇ From these data, what is the principal cellular mechanism of adipose tissue expansion in young adult guinea-pigs?

◆ Adipocyte enlargement. A fivefold increase in adipocyte volume (vertical axis) corresponds to a tenfold increase in the proportion of adipose tissue in the body (horizontal axis).

Scientists were impressed by the ability of adipocytes to undergo such large changes in volume. As a result of this finding, research on growth of adipose tissue has been directed mainly towards studying the mechanism of adipocyte enlargement and the factors that control adipocyte volume. However, these data show that increases in the numbers of cells also contributes to expansion of the tissue: if a fivefold increase in adipocyte volume corresponds to a tenfold increase in adipose tissue mass, the cells must be about twice as abundant (in proportion to lean body mass) in the fattest specimens as in the leanest specimens.

The total mass of adipose tissue of humans cannot normally be measured directly in the same way as is possible for guinea-pigs but the proportion of adipose tissue in living people can be estimated approximately using several indirect methods, such as measuring body density by weighing the person in air and under water. Ninety per cent of the mass of adipose tissue is lipid, which makes it significantly less dense than watery tissues such as muscle or guts, or hard tissues such as bone. Adipocyte volume can be measured from small samples of adipose tissue removed during surgical operations or using a biopsy needle. Figure 3.10a shows the relationship between mean adipocyte volume and the total mass of adipose tissue for 279 Swedish adults aged 22–58 years, about half of them people attending hospital clinics. The correlation is clearly not as close as it is for the data in Figure 3.9.

◇ Can you suggest some reasons why the human data may be more scattered than those from guinea-pigs?

◆ The human subjects were of a much wider range of ages; they probably had much more varied diets and exercise habits than the guinea-pigs; some of the subjects recruited from clinics may have been suffering from diseases that affect adipose tissue.

In other words, the 'experimental conditions' under which the humans were living were not as tightly controlled as those in which the guinea-pigs were maintained. The scientists have to decide whether these factors are sufficient grounds for dismissing the weak correlation between body composition and adipocyte volume in Figure 3.10a as spurious, and concluding that the mechanisms of adipose tissue growth are essentially similar in humans and guinea-pigs, or whether to conclude that the data show that the mechanisms of adipose tissue growth are fundamentally different in these two species. The two species belong to different orders of mammals (see the Appendix): humans are primates and guinea-pigs are rodents.

◇ How could comparative studies help to resolve this problem?

◆ Making similar measurements on other primates and other rodents would shed some light on the issue. If similar data from other primates resemble that from humans but differ from that from all other rodents, the contrasts between Figures 3.9 and 3.10 probably arise from fundamental differences between taxonomic groups.

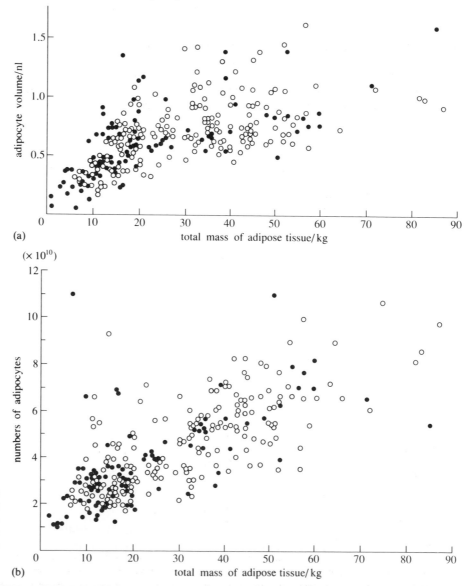

Figure 3.10 Relationship between the cellular structure of adipose tissue and its total mass in 279 adult humans. (a) Mean adipocyte volume and the mass of adipose tissue. (b) Estimated total adipocyte number and the mass of adipose tissue. Men: black circles; women: white circles.

Figures 3.11a and b show data similar to those in Figures 3.10a and b from crab-eating macaque monkeys (sometimes called cynomolgus monkeys) that had been bred and maintained in capitivity.

◇ What is the principal cellular mechanism of expansion of adipose tissue in these monkeys?

◆ Increase in the numbers of adipocytes.

The number of adipocytes per kilogram of lean body mass increases more than twelvefold over a fifteenfold range of fatness (Figure 3.11b) but, above 5% fatness, there is no significant correlation between adipocyte volume and the proportion of adipose tissue in the body (Figure 3.11a). However, it is still not clear whether the apparent increase in numbers of mature, readily recognizable adipocytes is due to the formation of new cells (by cell division), or the maturation of pre-existing cells that were too small to be noticed in the leaner specimens.

◇ From Figure 3.10b, is there any evidence for increase in the numbers of adipocytes in humans?

◆ Yes. Above about 15 kg of adipose tissue, adipocyte volume does not correlate with fatness (Figure 3.10a), but the numbers of adipocytes continue to increase over the whole range of measurements (Figure 3.10b), so increase in cell numbers must make a substantial contribution to the sixfold increase in adipose tissue mass observed in this sample of humans.

Humans appear to resemble monkeys much more closely than rodents. The uncomfortable conclusion is that research into the factors that control adipocyte volume in laboratory rodents may be irrelevant to understanding the mechanism of expansion of adipose tissue in primates, including humans. Instead of concentrating on adipocyte enlargement, it would have been much more useful to have studied the factors that control increases in the numbers of mature adipocytes.

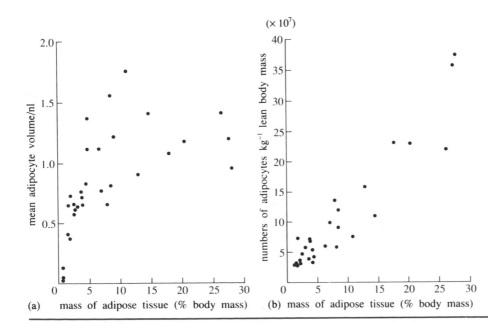

Figure 3.11 Relationship between the cellular structure of adipose tissue and fatness in 28 adult *Macaca* monkeys born and maintained in captivity. (a) Relationship between mean adipocyte volume and fatness (mass of adipose tissue as a percentage of the total body mass). (b) Relationship between number of adipocytes per kg lean body mass and fatness.

◇ Would young adult guinea-pigs be suitable animal models in which to study this phenomenon?

◆ No. The data in Figure 3.9 show that the number of adipocytes in young adult guinea-pigs increases at most twofold.

Unfortunately, the adipose tissue of young rats, hamsters and other laboratory rodents seems to grow in the same way as that of young guinea-pigs. Studies of these species are therefore unlikely to reveal very much about the mechanism of adipocyte proliferation that seems to contribute so much to adipose tissue growth in humans and other primates.

The data in Figure 3.9 from guinea-pigs more than 14 months old suggest that older specimens would be more promising as animal models of adipocyte proliferation: mean adipocyte volume remained low in some of the fattest specimens, indicating that there must be a substantial increase in the number of adipocytes in these individuals. However, most scientists are reluctant to meet the additional expense of keeping the animals until they are older. Furthermore, without major manipulation of the diet or husbandry conditions, even older guinea-pigs (maximum recorded fatness: 19% of the body mass is adipose tissue; Figure 3.9) do not become as obese as monkeys (maximum recorded fatness: 28%) or humans (maximum recorded fatness: more than 50%).

3.3.2 Natural animal models

Some biological questions cannot be answered by experimentation alone. In most mammals, it is difficult to tell males from females except by examining the genitalia, but we can distinguish men and women at a glance from the shape and proportions of their bodies. From infancy to puberty, boys and girls are quite similar (apart from the external genitalia), but, from adolescence, sex differences in body size and body shape become conspicuous.

◇ Do the data in Figure 3.10 show any sex differences in body composition?

◆ Yes. The women are fatter. There is much variation between individuals, but the average quantity of adipose tissue is slightly greater in women. Since women are on average shorter, and have less massive skeletons and muscles than men, a greater proportion of their bodies is adipose tissue.

As well as differences in the abundance of adipose tissue, there are also sex differences in its distribution on the body. Much of the breast, and a significant proportion of the leg, upper arm and abdomen are adipose tissue and, in adults, sex differences in body conformation are due mainly to the size and shape of the skeleton and the attached muscles, and to the quantity and arrangement of adipose tissue. Women are also normally fatter and, since proportionately more of their adipose tissue is superficial, it is a conspicuous feature of their bodies. Biologists have long been interested in the development, physiological implications and evolutionary origins of this prominent feature of human anatomy. One of the most widely accepted theories about the evolutionary origins and natural function of adipose tissue proposes that certain depots are used selectively to meet the energy needs of pregnancy and lactation and that these specialized energy stores are therefore larger in women. At first glance, this idea seems reasonable enough and there is some evidence that supports it: women's fat stores are indeed depleted if they are undernourished while reproducing; also during lactation, certain enzymes that promote lipid breakdown become more active in some adipose depots, particularly the thigh, but not in other typically feminine sites, such as the breast. However, when this explanation is considered from a comparative point of view, it seems less plausible.

Most suckling mammals can digest a lipid-rich diet and indeed seem to obtain almost all their energy from breakdown of fats, even if they mainly utilize carbohydrates such as glucose when adults. Carbohydrates from the diet can be converted into fatty acids but not the reverse, so the carbohydrate in milk cannot be synthesized from the fatty acids in storage lipids. Table 3.1 shows the average fat, protein and carbohydrate contents of milk from various

Table 3.1 The average lipid, protein and carbohydrate content (% total fresh weight) of milk from various mammals.

Order and species	Lipid	Protein	Carbohydrate
Primates			
Baboon	4.6	1.5	7.7
Orang-utan	3.5	1.4	6.0
Chimpanzee	3.7	1.2	7.0
Woman	4.1	0.8	6.8
Rodentia			
Guinea-pig	5.7	6.3	1.7
House mouse	13.1	9.0	3.0
Beaver	19.0	11.2	2.2
Brown rat	8.8	8.1	3.8
Carnivora			
Fox	5.8	6.7	4.6
Dog	9.5	7.5	3.8
Mink	7.3	5.6	4.5
Cat	10.8	10.6	3.7
Himalayan bear	10.8	8.4	3.2
Brown bear	23.0	11.1	0.6
Polar bear	33.0	10.9	0.6
Pinnipedia			
Northern fur seal	49.4	10.2	0.1
Harp seal	42.2	8.7	0.1
Californian sea-lion	30.7	8.6	0.3
Proboscidea			
African elephant	5.0	4.0	5.3
Indian elephant	7.3	4.5	5.2
Perissodactyla			
Rhinoceros	0.2	1.4	6.1
Donkey	0.6	1.4	6.1
Horse	1.3	1.9	6.9
Artiodactyla			
Pig	8.3	5.6	5.0
Giraffe	4.8	4.0	4.9
Cow	3.7	3.2	4.6
Water buffalo	6.5	4.3	4.9
Gazelle	8.8	8.8	5.7
Sheep	7.3	4.1	5.0
Moose	10.0	8.4	3.0
Red deer	8.5	7.1	4.5
Reindeer	10.9	9.5	3.4

species of mammals. The composition varies greatly between species (and sometimes according to the age of the suckling offspring) depending upon the duration of lactation, the availability of food for the mother and many other factors.

◇ Does the composition of human milk suggest that storage lipids contribute more to its synthesis than is the case in most other mammals?

◆ No. Compared to the milk of most other mammals, human milk, like that of other primates, is low in fat but rich in carbohydrate.

The composition of human milk suggests that lipids in adipose tissue could not contribute much to its synthesis. The carbohydrates and proteins must be derived mainly from the diet and indeed our knowledge of the ecology and physiology of primitive peoples suggests that women (like most other female mammals) normally eat regularly throughout pregnancy and lactation. Nearly all mammals breed seasonally and the birth of the young normally coincides with ready availability of large quantities of food. Lipid and protein stores are a supplement and emergency reserve that maintains the milk supply in those species, particularly large predators, for which the food supply may be erratic, with long, irregular intervals between meals.

Polar bears (and some species of seals and whales) are exceptions: the young are born far away from the mother's principal source of food, and she does not feed for several months before and after giving birth. Polar bears prey on seals; they prey on newborn seals during the brief period when they are on ice-floes or beaches, but they catch the adults mainly as the seals surface to breathe through holes in the ice. Hunting is successful only when the sea is almost completely frozen over, so, except for eating a few berries and seaweed in the summer, polar bears of both sexes fast from late June until the sea freezes again in October. The males and non-breeding females feed throughout the winter, but the pregnant females migrate inland to areas where large snowdrifts form. They dig out a maternity den in the snow, in which they give birth and suckle their litter for three to four months without feeding themselves. Newly born polar bears are only about the size of a large rat (body mass about 500 g), but they grow quickly and weigh about 10 kg by the time they emerge from the maternity den.

◇ From Table 3.1, how does polar bear milk differ from milks of other mammals?

◆ It is very rich in lipids, above average in protein content but low in carbohydrates.

Only pinnipeds produce milk that contains about as much or more lipid than polar bear milk.

◇ Can the composition of polar bear milk be attributed to the bears' diet?

◆ Possibly, because their principal food, seals, also produce very rich milk and so probably have massive lipid stores.

However, the composition of milks of carnivores in general is not very different from that of herbivores such as rodents and artiodactyls, although those of large herbivores such as rhinos, horses and elephants contain much less fat. The milk of the mainly herbivorous brown bears (see Chapter 2, Section 2.1.1) is also richer in lipids than that of predatory carnivores, such as cats, mink and foxes, so diet cannot be the only factor.

All the nutrients that the mothers use to sustain themselves and their rapidly growing offspring are sequestered in the body stores several months before they retire to the maternity den. Much of these energy reserves are probably in the adipose tissue. Female polar bears are *adapted* to lactating during fasting. If sex differences in the distribution of adipose tissue are related to storage and release of nutrients during lactation, they should be pronounced in polar bears.

Figure 3.12 shows some data on the quantities of adipose tissue in superficial, intra-abdominal and intermuscular depots in adult and subadult polar bears.

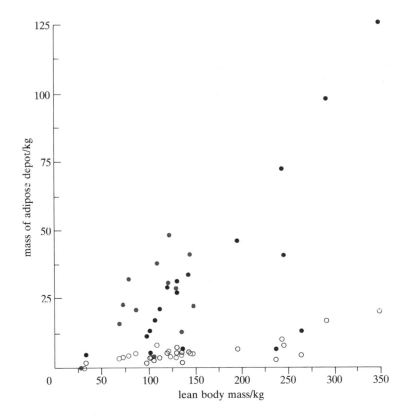

Figure 3.12 The mass of adipose tissue in intra-abdominal (open circles) and superficial (solid circles) depots in wild polar bears. Males: black symbols; females: red symbols.

◇ Is there any evidence for sex differences in the distribution of adipose tissue between these sites?

◆ No. The only difference is that the males are generally larger than the females.

There is no indication of an association between sex differences in the distribution of adipose tissue and energy storage during reproduction, even in a species that produces large quantities of lipid-rich milk while fasting. These data and similar measurements from other wild mammals do not support the theory that sex differences in the distribution of adipose tissue are essential for lactation. So we must explore other explanations for the presence of such conspicuous sex differences in the arrangement of adipose tissue in humans; among the possibilities under investigation are that the bulges of adipose tissue indicate social and sexual status, in much the same way as the distribution and colour of body hair enables us to estimate people's age and sex.

3.3.3 Animal models of disease

The study of the mechanisms of disease and the development and testing of remedies are some of the longest established and most important applications of biology. One of the most common causes of disease is the presence of pathogens: organisms, including viruses, bacteria, protistans, fungi, animals and (rarely) plants, that live parasitically in or on the host. Pathogenic organisms cause disease in many different ways, among them depletion of the host's food (e.g. gut parasites), release of toxic metabolites that interfere with the host's normal physiological processes, or by mechanical disturbance (e.g. blockage or damage to the gut or blood vessels). Although some drugs, such as morphine, were discovered to be effective against human ailments by chance, often by traditional healers, their specificity and effectiveness against the pathogenic organism or abnormal physiological process can nearly always be improved by modern research methods. Purification, minor modification of the chemical structure or changes in the means of administration can transform a crude remedy with numerous unpleasant side-effects into an efficient, safe medicine. Much of such research requires simulating the human disease in a laboratory animal.

Many pathogens, particularly animal parasites (see Chapter 2, Section 2.4.2), have complex life cycles involving several very different hosts. In order to study its physiology and life cycle, the pathogen must be maintained in the laboratory in an artificial culture medium containing nutrients and other essentials, or in suitable host organisms. Obviously the ideal host would be the species in which the pathogen is most virulent, usually those in which it occurs naturally. However, this choice is not always practical, particularly in the cases of human disease, or those of large, slow-growing plants or animals. The alternative is to find a plant or animal model, an organism in which the pathogen thrives and which can be more conveniently studied in the laboratory. This essential first step is rarely straightforward.

◇ Why would some pathogenic organisms fail to thrive when introduced artificially into laboratory animals?

◆ Pathogenic organisms are successful parasites that have anatomical and biochemical adaptations to living in or on their particular host, such as mechanisms that enable them to obtain all necessary nutrients from it, and to evade its defences.

Such adaptations are often elaborate and may have evolved following prolonged ecological association (Chapter 2, Section 2.4.2). Not surprisingly, many parasites do not survive if introduced artificially into a host that they never invade naturally. They also often have unusual nutritional requirements that make them difficult to maintain in culture under laboratory conditions.

As you might expect, pathogenic organisms are more likely to prosper in species closely related to their natural host. Ectoparasites such as fleas and lice (Chapter 2, Section 2.4.2) and many of the platyhelminth and nematode endoparasites that occur on gorillas, chimpanzees and other apes also infect humans (and vice versa). These primates are also susceptible to certain human bacterial diseases, notably tuberculosis, and a few viral diseases such as measles, a fact of increasing importance in the conservation of wild apes that come into more and more frequent contact with humans.

Some bacteria, such as those that cause sepsis in wounds, can live in a wide range of homoiothermic animals (although usually not in invertebrates or poikilothermic vertebrates). Pneumococci, the bacteria that cause pneumonia, invade the same tissues and cause the same symptoms in laboratory mice as they do in humans. They also proliferate readily in laboratory cultures if provided with suitable nutrients and maintained at mammalian body temperature (37 °C). It is therefore possible to study in detail the mechanisms by which these bacteria invade and injure the mammalian tissues and to devise remedies that eliminate the pathogens and restore normal function in the host's tissues. However, many other pathogenic bacteria, notably those that cause leprosy and syphilis in humans, have very specific and unusual nutritional requirements and have proved very difficult to maintain in any laboratory animal and almost impossible to culture artificially.

The rabies virus multiplies in a wide range of mammals (see Chapter 1, Section 1.1), but it is unusual in this respect: most of the viruses that cause disease in humans and domestic livestock do not multiply in the cells of convenient laboratory animals such as rats, mice and guinea-pigs. Certain species of monkeys are the only known organisms in which the virus that causes polio in humans will replicate; cynomolgus monkeys (Section 3.3.1) were used extensively in the 1940s and 1950s for the research which produced the polio vaccine that now protects children all over the world from this crippling disease. Most of the viruses responsible for the common cold (more than 100 different forms have been described) multiply only in humans and, for reasons that are not clear, in mink, a small carnivore that occurs wild in northern Europe and Asia and is now bred in captivity for its fur. Even when the pathogenic organism replicates in a non-natural host, its presence does not necessarily produce similar symptoms. The HIV virus (see Chapter 1, Section 1.1) only multiplies readily in primates and, less efficiently, in a few other mammals; it produces the symptoms of the human disease AIDS only in apes. Until they can manipulate the virus so that it multiplies readily in rats or other common laboratory species, or in cultured cells, biologists have to obtain supplies of the virus directly from human patients, or from captive primates that they have infected deliberately.

Until enough is known about the basis for the specificity of parasites and pathogens, the use of animals will continue to be indispensible for studying the mechanism of many diseases, and to develop and test remedies for them. But when studying diseases caused by micro-organisms and parasites, the species used as the model host must be carefully chosen to ensure that the life cycle and metabolism of the pathogenic organism are as similar as possible to those in its natural host. However, it should be emphasized that scientists have been able to develop drugs that combat most diseases caused by micro-organisms and animal and fungal parasites *because* there are animal models of the human conditions. Many other common and distressing human disorders, among them migraine headaches and morning sickness in pregnant women, have proved impossible to reproduce in laboratory animals. The lack of opportunity for studying these conditions experimentally is one of the main reasons for the delay in developing safe, effective remedies.

Summary of Section 3.3

Most experimental research into normal and abnormal physiological mechanisms involves using animal models. In some cases, the species used as models are chosen for practical reasons rather than because they are biologically similar to humans and other species which, for practical or ethical reasons, are more difficult to study experimentally. Comparison of similar physiological processes in a range of species helps biologists to assess the relevance of data from animal models to the other species. Disproportionate expansion of adipose tissue occurs in many captive and some wild mammals. In young laboratory rodents, the principal mechanism of adipose tissue expansion is adipocyte enlargement, but in captive monkeys and contemporary European humans, the same phenomenon is due mainly to increase in the numbers of mature adipocytes. Therefore, rodents are less appropriate than primates as animal models for research on this aspect of adipose tissue expansion as applied to humans. Comparison between natural obesity in wild animals with that of humans can support or refute theories about the origin of phenomena such as sex differences. The experimental study of diseases caused by pathogenic organisms involves finding a species in which the pathogen multiplies readily and produces effects similar to those that occur in their normal host species.

Question 3 (*Objective 3.4*) Answer the following questions in a few sentences:

(a) How does the study of spontaneous obesity in monkeys help scientists to assess the relevance to human obesity of experiments involving laboratory rodents?

(b) How does the study of wild polar bears help scientists to assess the validity of theories proposed to explain sex differences in the distribution of adipose tissue in adult humans?

Question 4 (*Objective 3.5*) Which of the following statements about disease-causing organisms are *true*?

(a) Pathogenic bacteria and parasitic animals reproduce in several species of hosts but viruses proliferate only in their natural host.

(b) Pathogenic organisms are more likely to thrive in host species that are taxonomically related.

(c) Parasitic organisms that are pathogenic in one host may survive in others but fail to produce recognizable symptoms of disease.

(d) It is rarely possible to predict the outcome of artificial introduction of a pathogen into an organism that is not its natural host.

3.4 HARNESSING BIOLOGICAL PROCESSES TO HUMAN NEEDS

There are many ways of exploiting the natural world for human use: the oldest are hunting and gathering, including chopping down trees for timber, fishing, harvesting mushrooms, fruits and other wild plants. Effective exploitation of the organisms depends only upon knowing where to find them and how to catch them. About 5 000 years ago, humans developed agriculture and began breeding and raising livestock in captivity. Such activities require an elementary knowledge of nutrition and animal husbandry, and the effects of grazing on wild plants. Crops and livestock were also made more productive or hardier by selective breeding, but until this century scientific biology played little part in achieving these objectives or determining the means of improving domestic animals and plants. In this section, we introduce you to some more subtle (and less destructive) economic uses of 'wild' organisms and show you how their successful exploitation for practical purposes is based upon knowledge of their physiology and biological relationships to other organisms.

3.4.1 Chemical defences of sedentary organisms

Almost all animals and fungi feed off the tissues, secretions or excretions of other animals, plants or micro-organisms (Chapter 1, Sections 1.5 and 1.6). With a few exceptions (see Chapter 2, Section 2.4), the 'prey' is damaged or destroyed by the predator, parasite or herbivore. Large 'prey' organisms, such as many adult insects, most fishes, birds, rabbits, deer and humans, avoid predation by flying, swimming or running away. Large, mobile 'predators' can move to another feeding area if a competing organism consumes the food source. However, such escape is impossible for micro-organisms, plants and the many kinds of animals which move slowly or not at all. Predation of such sedentary organisms and competition between them for food are minimized by the presence of structures or chemicals that render them inedible, distasteful or even poisonous.

◇ Describe some ways in which sessile or slow-moving animals are protected from predation.

◆ Snails (Plate 13b), mussels, barnacles (Plate 11h) and tortoises (Plate 15g) burrow or hide from predators, but they also have strong shells; the spines of some sea-urchins (echinoids) are very sharp and/or contain toxins (Chapter 1, Section 1.6.6 and Plate 14a).

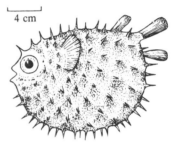

4 cm

Puffer fish (Order Tetraodontiformes) occur in brightly lit shallow seas and around coral reefs but are clearly not streamlined for fast swimming; indeed, they swim slowly and weakly, paddling themselves along with their fins and stumpy tail. Some species of puffer fishes such as *Tetraodon* (Figure 3.4) contain lethal toxins, but other closely related species, such as the porcupine fish, *Diodon hystrix* (Figure 3.13), are non-poisonous. They have stout spines and when attacked, swallow large quantities of water that inflates the body and protrudes the spines, making them almost impossible for predators to swallow. *Phyllobates* (Section 3.2.1) spends most of the time sitting on trees and other conspicuous places, from where it ambushes passing insects. But for their poisonous and foul-tasting toxins or spines, these animals would be vulnerable to predation. Such mechanical and chemical defences occur naturally in many sedentary organisms, particularly micro-organisms and plants, as protection from predation and herbivory.

Figure 3.13 Diodon hystrix (porcupine fish), a non-poisonous species of the order Tetraodontiformes. When undisturbed, the body is more flattened and the spines are laid back, and they swim slowly using the tail and pectoral fins. But when attacked, as in this drawing, porcupine fish puff themselves up, protruding their stout spines, which makes them almost impossible for predators to swallow.

◇ Which group of animals includes the greatest number and diversity of herbivores?

◆ Insects (see Chapter 1, Section 1.6.4 and Chapter 2, Section 2.4.3).

The insecticidal properties of pyrethrum have already been described (see Section 3.2.3); its natural function in the African daisy is protection of the foliage from herbivorous insects. In fact, protection is rarely complete, because many herbivorous animals that can tolerate small quantities without ill-effect may poison themselves if they eat large quantities of foliage that contains such toxins. Shoots and young leaves are often softer and more digestible because they contain a small proportion of toxins, and large, mobile herbivores such as monkeys, move from tree to tree, taking only a few leaves or buds from each plant. However, mature foliage from a single tree may be the only food available to small invertebrates such as caterpillars and so the presence of even small quantities of noxious substance may be sufficient to prevent serious damage to the plant. A huge variety of substances, many of them chemically elaborate, occur in the leaves, seeds and bark of angiosperm plants and protect them against herbivory. Many of them, including caffeine (in tea, coffee and cola nuts), nicotine (in tobacco), morphine (in poppy seeds), salicylic acid, the principal ingredient of aspirin (in willows), cocaine, camphor and curare (see Section 3.2.2) also modify biochemical processes in humans, and have been the sources of some of our most useful drugs. Understanding of both the natural mechanisms of synthesis and the biological basis for its occurrence can assist efficient exploitation of this abundant and diverse source of valuable natural products.

3.4.2 Human uses for natural defences

Fungi (see Chapter 1, Section 1.5) and bacteria (see Chapter 1, Section 1.2.1) are among the most abundant and effective consumers of excrement and dead plants and animals on land (Plate 8a), but many other kinds of organisms, including insects such as termites and the maggot larva of the housefly (see Chapter 1, Section 1.6.4) also feed on such food. Many fungi and micro-organisms synthesize a wide range of substances that may serve to deter other bacteria, protistans and scavenging animals from consuming the fungus and its food source. Humans and many other animals can smell or taste many of the defensive chemicals, which may remain in the organisms or be secreted. A few, such as truffles, bakers' and brewers' yeast and the moulds that are used to make cheese, produce desirable flavours, and have been important as food for humans for thousands of years. Many other fungi and micro-organisms are distasteful to humans and probably also to some other animals. Fortunately, most of the fungi that are toxic to humans also smell or taste unpleasant, but the most common of the poisonous species in Britain, the deathcap (*Amanita phalloides*), is highly dangerous because it is very similar in appearance to a harmless species and, more significantly, because it is palatable.

Bacteria (Chapter 1, Section 1.2.1) are a major cause of disease in humans and other vertebrates. Many different kinds also feed on dead plants and animals, among them actinomycetes (Figure A.1) that form long chains of cells that superficially resemble fungal moulds. If actinomycetes are introduced onto the same food source as other bacteria, multiplication of the latter is curtailed and they may stop entirely. A greater proportion of the food source becomes available to the actinomycetes because they produce 'something' that eliminates many of the other kinds of bacteria. Other organisms, notably the ascomycete fungus *Penicillium* (Figures 1.15c and A.11a) secrete similar anti-bacterial substances, called antibiotics. The prokaryotic actinomycetes or eukaryotic fungi can be grown artificially, often on agar jelly enriched with nutrients, and the antibiotics that they secrete can be extracted from the culture medium and administered to humans and domestic livestock. Proliferation of many, but unfortunately not all, disease-causing bacteria is curtailed by antibiotics.

◇ Would antibiotics be toxic to vertebrates?

◆ Not if the biochemical processes that they interfere with in prokaryotes are absent or substantially different in vertebrates.

In many cases, there are sufficient differences between the pathogen and the host for the former to be eliminated with minimal interference with the host's metabolism. However, most antibiotics are small molecules that permeate most kinds of cells, and such is the complexity of the biochemical pathways that there are sometimes unexpected side-effects, such as damage to the ear and delicate parts of the nervous system.

It would be quite wrong to leave you with the impression that scientists developed antibiotics by applying the simple biological reasoning just described. On the contrary, although it has been known for at least two thousand years that open wounds heal faster and more completely if maggots are allowed to live on them (blowflies were sometimes introduced deliberately for this purpose), understanding of the natural biological interactions between organisms was not sufficiently advanced for people to ask why carrion-feeding larvae should have this unexpected property, or try to isolate the active ingredient. Penicillin was not discovered until the late 1920s, and not developed for large-scale medical use until the 1940s, more than fifty years after bacteria had been shown to be a major cause of illness and poor wound healing (see Chapter 1, Section 1.2.1). The basis for the beneficial effects of maggots proved to be an anti-bacterial substance that they secrete. Many people might have been saved from debilitating and often lethal infections if scientists had realized that good organisms in which to look for a potent anti-bacterial agent were those that naturally interact with bacteria.

◇ Are there other organisms that destroy bacteria selectively in a way that would be less likely to cause side-effects in higher animals?

◆ Bacteriophages (Chapter 1, Section 1.1) are viruses that parasitize, and often kill, bacteria by incorporating themselves into their genetic material and disrupting the metabolism of the cell. Normally, a single kind of phage attacks only one or a few closely related species of bacteria.

Phages of certain bacterial decomposers are already cultured artificially and applied to silage, where they selectively eliminate the bacteria that render the food unpalatable to cattle, while being harmless to those that produce satisfactory silage. But bacteriophages are not yet used widely to destroy pathogenic bacteria in humans and livestock.

◇ Would antibiotics be effective against protistan diseases and parasitic worms?

◆ Probably not, particularly if they produce few 'side-effects'. Protistans and parasitic worms, like vertebrates, are eukaryotes, which are more similar to each other than any of them is to prokaryotes (see Chapter 1, Sections 1.2 and 1.3).

Side-effects of antibiotics arise because the drug interferes with eukaryotic biochemical processes, as well as curtailing the activities of the prokaryotic pathogens. If the drug produces few side-effects in the animals or people it is unlikely to interfere with the metabolism of eukaryote organisms sufficiently to be an effective means of eliminating pathogenic protistans or worms. You can see why diseases caused by protistans, e.g. malaria (Chapter 1, Section 1.3.1) and sleeping sickness (Figure A.15a), and infestations of parasitic worms, e.g. schistosomiasis (Chapter 2, Section 2.4.2; Plates 10(b)–(f) are difficult to eradicate completely and safely using drugs. *Cinchona* is a large tree native to the Amazonian rainforest, the bark of which contains quinine, a complex organic compound that inactivates the protistans *Plasmodium* spp. that cause malaria in humans. Quinine tastes very bitter to humans, and probably also to other herbivores, so its natural function may be protection of the plant from herbivory. It is not clear why quinine is toxic to *Plasmodium vivax* since the two organisms do not come into contact in the wild. The Aztec Indians were using a crude extract of the bark as a cure for malaria long before Columbus first visited the Americas. Unfortunately, the quantities of quinine vary greatly between specimens, with the bark of some trees containing as little as 4% or as much as 13% dry weight of the drug. Since they could not isolate quinine or measure its concentration in the bark accurately, this variation between sources made it very difficult to administer the appropriate dose.

◇ Can you suggest why the concentration of quinine in the bark should be so variable?

◆ If its natural function is to act as a defence against herbivory, the type and abundance of herbivores attacking the trees may be associated with genetic differences between populations of the species, or individual plants may be able to synthesize more or less quinine in response to local conditions.

Unfortunately, like many rainforest trees, *Cinchona* is relatively rare and flowers irregularly and sometimes as infrequently as once in five years. The biologist and explorer, Charles Ledger (1818–1905) went to great lengths to find out from Amazonian Indians which *Cinchona* tree produced the best quality quinine. Seeds for the first *Cinchona* plantation were obtained by camping for several years under that tree, waiting for it to flower and set seed. Because the demand for quinine far outweighs supplies, even from cultivated trees in India and Indonesia, most anti-malarial drugs in use today are synthetic, but many of them are chemically similar to natural quinine.

Cells in captivity

Keeping a little frog in captivity does not take much space; keeping kraits (Section 3.2.2) or puffer fish (Section 3.2.1) involves more elaborate apparatus and all these animals have unusual and specific food requirements. However, maintaining large forest trees 'in captivity' is virtually impossible. They often survive only under very specific conditions and are susceptible to disease when planted at high density; many species grow quite slowly, so they may be a century old before yielding large quantities of the required product. Obtaining material from wild specimens is not only potentially harmful to the plants, it is expensive and inefficient; as much as 200 kg of leaves may be needed to extract 1 g of a useful drug. With the rapid destruction of natural vegetation, particularly tropical forests, wild specimens are becoming rare.

The solution may be to culture cells isolated from the mature plant in the laboratory. In theory, the cells could be propagated indefinitely, and batches harvested to extract the useful product. The method has worked reasonably well for a few plants, notably *Cinchona* and the chilli-pepper, *Capsicum frutescens*, from which the spice capsiacin is extracted. However, in many cases the yield of the spice is lower than it would be from an equivalent mass of cells in the intact plant.

◇ Using your knowledge of the natural role of such substances in the wild plants, can you suggest a way of prompting the cells to synthesize more of the desired product?

◆ If the spice acts as chemical protection against herbivores or micro-organisms, simulating the presence of its natural attackers may induce higher production.

In certain cases, adding the extract of the fungus that feeds on the species in the wild can increase the rate of synthesis of natural fungicides in the cultured plant cells.

Our knowledge of the natural mechanism of sexual reproduction and infection of bacteria by viruses is now sufficiently advanced that biologists can control their biosynthetic activity directly by altering their genetic material. In this way, the biochemical machinery of bacteria is harnessed to synthesize large, complex molecules that cannot be manufactured by ordinary chemical methods. Biologists can insert into the bacterial chromosome lengths of DNA 'genes' that code for the synthesis of particular proteins. If maintained under suitable conditions and supplied with the right nutrients the bacteria synthesize and secrete substances 'to order'. The required product is then extracted from the bacterial culture and purified using standard chemical methods. At present, certain drugs that are normal components of plants and animals, particularly hormones and other small 'messenger' molecules such as interleukin and interferon, and even structural materials such as tooth enamel can be produced by bacteria. During the last ten years, biologists have perfected methods for culturing actinomycetes and many other bacteria on a large scale, and for modifying their genes so that they synthesize large quantities of medically useful antibiotics.

3.4.3 Exploiting natural biochemical mechanisms

Diseases such as cancer, arthritis, atherosclerosis and senile dementia present even greater problems than infectious diseases because they arise largely (in many cases probably entirely) within the animal itself and other organisms or externally generated toxins are not involved. The diseased tissues are chemically so similar to the rest of the body that it is almost impossible to destroy the defective cells without damage to healthy cells, particularly rapidly dividing cells, such as hair and the lining of the gut.

Extracts from wild plants are providing new ideas for drugs that destroy tissues affected by such diseases with minimal side-effects. The seeds of the common mistletoe and of the castor oil plant, a handsome shrub widely cultivated in gardens in warm countries, contain ricin, a poison so effective that chewing as few as four seeds produces serious illness in adult humans, and a dose of more than eight seeds is invariably fatal. Recently, biologists have studied exactly how ricin is so toxic in such small quantities. Ricin consists of two linked polypeptides (the A- and B-chains), each of which is responsible for a different step in the poisoning process. Nearly all metabolically active cells have on their surface a variety of receptors, consisting of one or several large protein or lipid molecules, that facilitate the uptake of harmless, useful metabolites from the extracellular fluid or the blood. The A-chain of ricin binds to certain such receptors, thereby enabling both chains to cross the cell membrane and enter the cell. The chains then separate (by the breakage of a single disulphide bond) and the B-chain binds to and permanently inactivates ribosomes.

◇ How would the presence of a protein that binds ribosomes affect an organism?

◆ Ribosomes are essential to protein synthesis, and almost all cells contain them.

In fact, ricin inactivates ribosomes so effectively that one molecule can disable the protein-synthesizing mechanisms of an entire cell, thereby preventing it from growing, dividing or maintaining secretion.

◇ What features of ricin might be of interest to cancer biologists?

◆ The A-chain offers a means of breaching the cell membrane which is normally impermeable to large molecules, while 'towing' a therapeutic molecule. The B-chain is an effective poison for actively growing cells that contain many ribosomes.

Research is underway to harness the impressive properties of this natural substance. Projects directed towards coupling ricin to molecules that bind specifically to cancer cells look particularly promising.

Summary of Section 3.4

The structural and biochemical diversity of organisms can be exploited to provide economically useful materials. Compounds synthesized by plants, micro-organisms and sedentary animals that protect them from predation are useful sources of insecticides, antibiotics and other substances that are toxic to a specific group of organisms. Once identified, the useful material can be extracted from wild organisms, but this approach is impractical for rare species. If the organisms can be successfully maintained in artificial conditions, they can often be manipulated, either by genetic modification or by stimulating the natural biosynthetic mechanisms, to synthesize greater quantities of the useful material.

Question 5 (*Objective 3.6*) Which of the statements (a)–(f) are true of the identification and exploitation of economically useful natural substances?

(a) Most useful drugs and pesticides are synthesized by structurally simpler organisms such as prokaryotes and fungi, but not by plants or animals.

(b) Organisms never produce substances that are toxic to members of the same phylum or division.

(c) It is possible to predict from its taxonomic affinities whether a species is a likely source of useful drugs or pesticides.

(d) It is possible to predict from its diet, habits and relationships to other organisms whether a species is a likely source of useful drugs or pesticides.

(e) Information about an organism's diet, habits and relationships to other organisms is helpful as a guide, but it is never an infallible indicator of whether it is a likely source of useful drugs or pesticides.

(f) A substance produced by one organism will be toxic to another organism only if the two species interact naturally in the wild.

3.5 CONCLUSION

These examples illustrate how features of the basic body plan, such as the size and anatomical relations of the major tissues, and specific adaptations, such as toxins that have evolved in predators or herbivores feeding on particular species, offer unique opportunities for elucidating basic physiological mechanisms, studying the mechanisms of infection and disease and producing economically useful materials. The study of the diversity of organisms is important for recognizing and interpreting such opportunities, as well as for explaining the origin and evolution of species and the relationships between them and their environment—and simply knowing more about the structure and habits of the great wealth of micro-organisms, plants and animals that share our planet.

OBJECTIVES FOR CHAPTER 3

Now that you have completed this chapter, you should be able to:

3.1 Define and use, or recognize, definitions and applications of each of the terms printed in **bold** in the text.

3.2 Outline the comparative and experimental methods of investigating biological problems. (*Question 1*)

3.3 Describe the advantages of using invertebrates and non-mammalian vertebrates for studying the basic mechanisms of neural transmission. (*Question 2*)

3.4 Explain how comparative studies are used to assess the relevance of studies in laboratory animals to normal and abnormal physiological processes in humans and domestic livestock. (*Question 3*)

3.5 Describe some problems associated with the use of animal models for studying diseases caused by pathogenic organisms. (*Question 4*)

3.6 Give some examples of applications of biological knowledge to the development of economically important processes and materials. (*Question 5*)

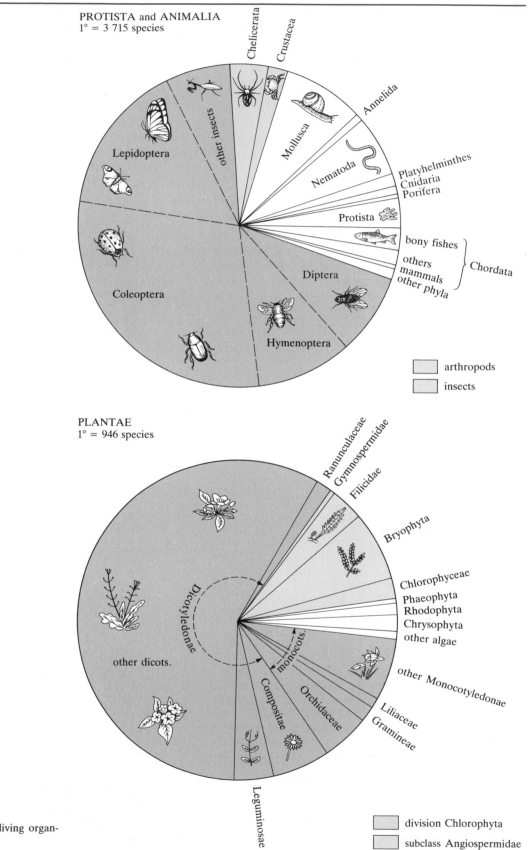

PROTISTA and ANIMALIA
1° = 3 715 species

Chelicerata
Crustacea
Annelida
Mollusca
Platyhelminthes
Cnidaria
Porifera
Nematoda
Protista
bony fishes
others
mammals
other phyla
} Chordata
other insects
Lepidoptera
Coleoptera
Diptera
Hymenoptera

arthropods
insects

PLANTAE
1° = 946 species

Ranunculaceae
Gymnospermidae
Filicidae
Bryophyta
Chlorophyceae
Phaeophyta
Rhodophyta
Chrysophyta
other algae
other Monocotyledonae
Liliaceae
Gramineae
Dicotyledonae
other dicots.
monocots.
Compositae
Orchidaceae
Leguminosae

division Chlorophyta
subclass Angiospermidae

The relative abundance of living organisms.

APPENDIX: A SURVEY OF LIVING ORGANISMS

INTRODUCTION

This Appendix is a systematic survey intended to provide basic information about groups of living organisms. It is written in a concise style and is intended as a work of reference for background information. It is not a comprehensive list and deals mainly with the groups mentioned in Chapters 1–3. Some taxonomists 'split' and others 'lump' groups: for example, grasshoppers, cockroaches and stick insects are put together in the insect order Orthoptera by 'lumpers' but assigned to three separate orders by 'splitters'. This classification involves moderate splitting. You are not expected to read straight through this survey but should consult it when you wish to know more about groups that are referred to in the main text or when you wish to remind yourself of the special features of any group.

Viruses (Figure 1.1)* are non-cellular organisms composed of protein and nucleic acids only. They reproduce only inside cells and are not classified in the same way as cellular organisms.

The prokaryote kingdoms Archaebacteria and Eubacteria are cellular organisms that have small ribosomes and may also contain cytoplasmic granules and vacuoles but lack membrane-bound organelles such as nuclei, mitochondria and chloroplasts. The eukaryote kingdoms Plantae, Fungi, Protista and Animalia include cellular organisms that have larger ribosomes and membrane-bound organelles, including nuclei, mitochondria and chloroplasts.

KINGDOM ARCHAEBACTERIA

Minute, mostly anaerobic prokaryotic organisms which differ fundamentally from Eubacteria and from eukaryotic organisms in the structure of their cell walls, the outer cell membrane and length and sequence of the ribosomal RNA, and the basic mechanism of transcription and translation of genetic material. Most Archaebacteria live in 'extreme' habitats such as hot sulphurous springs, stagnant swamps and salty seas and may be similar to very ancient organisms.

*Figures in the main text are referred to in this way; figures in this Appendix are numbered A.1 etc.

KINGDOM EUBACTERIA

Bacteria are 0.5 to 5 µm (Figure 1.2), with rigid cell walls mainly composed of mucopeptides (amino acids plus sugar derivatives), sometimes with lipopolysaccharides (fats plus sugar derivatives). The cells may be of different shapes and may form loose associations or chains (Figure A.1). Many bacteria have one or more flagella, each consisting of a single fibril composed of a single protein, flagellin. Vacuoles are sometimes present. There is no nucleus; genetic material is one or more circular molecules of DNA that are free within the cell. Reproduction is normally by fission (a cell splits into two) but some have a sexual process. Some species form thick-walled, dormant spores.

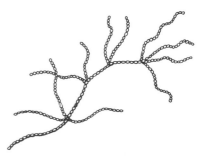

Figure A.1 Streptomyces, an actinomycete bacterium.

Bacteria may be free-living organisms, parasites or mutualistic symbionts; some are pathogenic to plants and animals. They have a great variety of structure and metabolic pathways and include photoautotrophs (anaerobes with a peculiar type of chlorophyll), chemoautotrophs (autotrophes that obtain energy from chemical compounds rather than sunlight; see Chapter 1, Section 1.2 and Chapter 2, Section 2.1), anaerobic heterotrophs (fermenters) and aerobic heterotrophs; some can fix nitrogen. One of the basic classificatory tests is whether the cells stain with the Gram stain; those which give a positive reaction lack lipopolysaccharides in their walls and are susceptible to penicillin and to sulpha-drugs. Gram-negative forms have lipopolysaccharides and are susceptible to streptomycin, but not to penicillin.

Actinomycetes, e.g. *Streptomyces* (Figure A.1), form branching colonies and are Gram-positive spore-formers. They are common in soil and water and break down cellulose and chitin. In some features, they resemble fungi (with which they were formerly classified) but they are prokaryotes.

Cyanobacteria (also called blue–green bacteria or blue–green algae) are typically aerobic photoautotrophs, producing oxygen; many can fix nitrogen either in special cells called heterocysts or in ordinary cells. They may be single cells, groups of cells (e.g. *Microcystis*; Figure A.2), or filaments (e.g. *Anabaena*, *Oscillatoria*; Figure A.2) and may contain gas vacuoles. Cells (Figure 1.3) are usually at least 0.5 μm in diameter, with rigid walls similar to those of other bacteria and are usually enclosed in a mucilaginous sheath that includes cellulose fibrils (Plate 1). Chlorophyll *a* is present on lamellae in the cytoplasm; red (phycoerythrin) and blue (phycocyanin) pigments and unique carotenoids are also present and the cells are usually bluish-green in colour but may be red or black. Flagella are never present but some cells can glide by extrusion of mucilage. Many storage materials are formed, including glycogen. There is no nucleus; the DNA is circular and cells divide by simple fission; filaments may reproduce by fragmentation. No sexual processes are known. Some can form spores. Cyanobacteria live in the sea, freshwater and damp terrestrial places; planktonic species may form 'blooms' under favourable conditions. Some species are symbionts, e.g. as part of lichens or with liverworts.

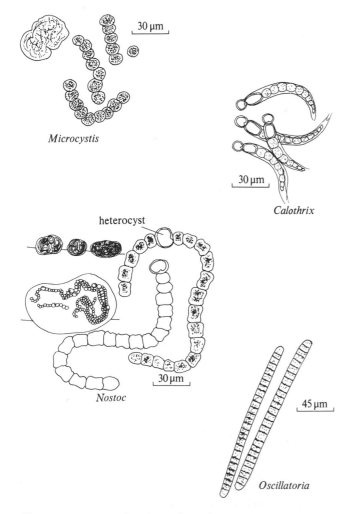

Figure A.2 Four species of cyanobacteria.

KINGDOM PLANTAE (about 350 000 species)

Plants are typically aerobic autotrophs, producing oxygen by photosynthesis in membrane-bound chloroplasts that contain chlorophyll *a* and *b*, β-carotene, xanthophylls and other pigments. All cells contain nuclei and the principal genetic material, nuclear DNA, is linear.

There are seven divisions, based on photosynthetic pigments, certain nuclear features and the biochemistry of cell walls and storage materials: six consist only of algae, the seventh includes certain algae and all other plants. Some classifications propose more divisions, and some algae, mainly motile unicellular or colonial forms, are sometimes included in the kingdom Protista.

Division EUGLENOPHYTA (about 450 species)

Motile, usually green algae, typically with one long and one very short flagellum that arise from a large gullet, e.g. *Euglena* (Plate 3a), *Peranema* (Figure 1.8a). Green species contain chlorophyll *b* in addition to *a*; the main storage material is a polysaccharide, paramylum. Some are colourless heterotrophs. They do not form colonies or filaments and live mainly in freshwaters. This division is sometimes classified in the kingdom Protista, subclass Phytomastigina, order Euglenoidida.

Division PYRROPHYTA (about 1 000 species)

Algae with a single, large nucleus; typically unicellular, with yellow–brown chloroplasts, forming starch, fats or oil as food reserve, and two unequal flagella, one in a transverse groove and the other in a longitudinal groove; the body is either naked or (usually) enclosed in cellulose plates, e.g. dinoflagellates (class Dinophyceae) (Figures 1.6a and b). The coloured species contain chlorophyll *c*, *a* and other pigments. Usually free-swimming in the marine and freshwater phytoplankton but some are colourless carnivores or parasites and others are coloured symbionts called zooxanthellae (e.g. in corals). This division is sometimes classified in the kingdom Protista, subclass Phytomastigina, order Dinoflagellida. *Ceratium* (Plate 3b), *Gymnodinium*, *Gonyaulax*, *Noctiluca* (Figures 1.6a and 1.6b).

Division CHRYSOPHYTA (about 6 000 species)

Mainly unicellular algae with walls of pectin plus some cellulose and a tendency to form external plates, usually of silica. The coloured species contain chlorophyll *a* and *c* and other pigments; the main storage materials are a polysaccharide, laminarin, and oils and fats. There are one or two flagella at some stage of their lives (sometimes only in reproductive stages).

Class Chrysophyceae

Golden algae: they include many small flagellated species (e.g. *Prymnesium*) and coccolithophorids that have small calcareous plates (these fossilized plates are abundant in chalk).

Class Bacillariophyceae

Diatoms (Figures 1.6c, d and e) are algal cells enclosed in a pair of siliceous box-like valves that are often elaborately 'decorated'; flagella are present only in a reproductive stage. Diatoms are very common in freshwater and marine plankton and are also attached to plants or the substratum. *Biddulphia*, *Chaetoceros*, *Navicula*, *Asterionella*.

Division XANTHOPHYTA (about 400 species)

Unicellular and colonial algae, usually yellow–green.

Some coloured species contain chlorophyll *e* in addition to *a*; the storage materials are oils and fats and sometimes a polysaccharide, leucosin. The cell walls are of cellulose plus pectin, sometimes including spicules of silica. Almost all species live in freshwaters; some form multinucleate filaments. Reproductive stages have two unequal flagella. *Vaucheria*, *Tribonema*.

Division PHAEOPHYTA (about 1 500 species)

Filamentous or thalloid multicellular algae, usually brown. The cells contain chlorophyll *c* in addition to *a* and a brown pigment, fucoxanthin; the main storage material is leucosin. The cell walls consist of cellulose plus pectin and alginic acids (mucopolysaccharides). Some reproductive stages have two unequal flagella. They are mainly marine, growing attached to rocks in the intertidal zone or in shallow seas. Some species form holdfasts, stipes and laminae and grow very large. Brown seaweeds: *Fucus* (Figure A.3a and Plate 4), *Ascophyllum*, *Laminaria* (Figure A.3b).

Division RHODOPHYTA (about 4 000 species)

Mainly marine non-motile algae without flagella at any stage and with cell walls consisting of cellulose plus pectin with polysulphate esters (from which agar is made), usually red. The coloured cells contain chlorophyll *a* and a red pigment, phycoerythrin; the main storage material is a form of starch. Rhodophyta include unicellular species, simple colonies, filaments and multicellular seaweeds with holdfast and lamina. Red seaweeds: *Gigartina*, *Polysiphonia*, *Porphyra*, *Lithothamnium*. There are a few freshwater forms, e.g. *Batrachospermum*.

Division CHLOROPHYTA (about 400 000 species)

About 7 000 species of algae and 330 000 species of 'higher plants', almost all green. The green cells contain chlorophyll *a* and *b* and the main storage material is starch.

Subdivision CHLOROPHYCOTINA (about 5 700 species)

Chlorophyta that do not form embryos; the green algae.

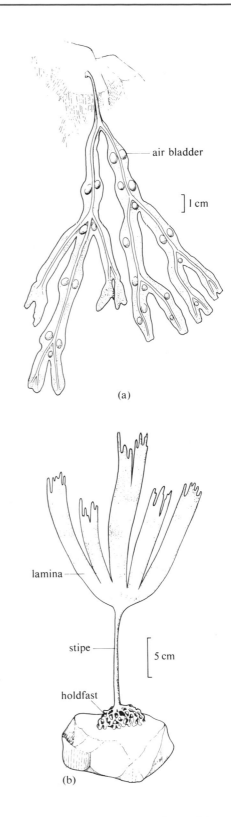

Figure A.3 Phaeophyta: (a) *Fucus vesiculosus*. (b) *Laminaria digitata*.

Class Charophyceae

A small class of freshwater, bottom-living algae. The thallus consists of an erect, branched axis with well-defined nodes and internodes, consisting of single, giant cells; whorls of laterals arise at each node. *Chara* (stonewort) (Figure A.4).

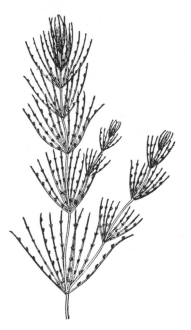

Figure A.4 Chlorophyta: Charophyceae, *Chara*. These algae form a calcareous skeleton which is gritty to the touch, hence their common name, stonewort.

Class Chlorophyceae

A very diverse group that include single cells, colonies, filaments (Plate 1) and some with a flat thallus plus holdfast, e.g. *Chlamydomonas*, *Volvox* (Plate 3d), *Spirogyra* (Plate 3c), *Ulva* (Figure 1.9) *Acetabularia*, *Zygnema* (Plate 1). Usually with two or four equal flagella and a single large chloroplast; no gullet; body wall usually cellulose; food reserve in green forms is starch. Mainly freshwater, some marine and terrestrial species. Some unicellular species are colourless (e.g. *Polytoma* (Figure 1.8b). The desmids (e.g. *Micrasterias*) are non-flagellate cells, constricted into two hemicells of characteristic shape, and are common in non-calcareous freshwaters. Some members of this class e.g. *Chlamydomonas*, *Volvox* (Plate 3d) are sometimes classified in the kingdom Protista, subclass Phytomastigina, order Phytomonadida.

Subdivision EMBRYOPHYTINA

Embryo-forming, mainly terrestrial plants with chloroplasts that contain chlorophyll *a* and *b* and with starch as the main storage material. All life cycles have an alternation of generations between (diploid) sporophytes and (haploid) gametophytes. The zygote forms the diploid embryo (a young multicellular organism) while still surrounded by the female parent's tissues. The egg (female gamete) is there-fore immobile and is not shed but the male gametes must be transferred in some way from male to female organs. The embryo grows into a sporophyte; after meiosis, the spores are formed and develop into haploid gametophytes which produce the eggs and male gametes and have varying degrees of independence from the sporophyte.

Superclass Bryophyta (about 23 600 species)

Liverworts, mosses and hornworts are low-growing plants with either a flat, shoot-like (thalloid) growth form or leafy shoots rhizoids are present but no true roots; the basic structure is relatively simple with no true vascular tissue. Growth points (meristems) are single apical cells. The gametophytes are the 'dominant' stage and bear the male and female reproductive organs: the sporophytes grow out of the fertilized egg cells and are often dependent on the gametophytes for nutrition (Figure 1.10). Water is essential for transfer of the flagellated male gametes.

Class Hepaticae (about 9 000 species)

Liverworts may be thalloid or leafy; many reproduce vege-tatively by shedding small groups of cells called gemmae from special pockets on the thallus. *Marchantia* (Plate 5a).

Class Musci (about 14 500 species)

Mosses are typically leafy and erect and have a more elaborate structure than liverworts. Some have a rudimentary vascular system. *Mnium* (Figure 1.10), *Poly-trichum*, *Tortula* (Plate 5b).

Class Antherocerotae (about 100 species)

Hornworts: the gametophytes are flat and thalloid; the sporophytes have continuous growth from a basal mul-ticellular meristem embedded in the gametophyte thallus.

Superclass Tracheophyta (vascular plants)

Xylem and phloem (vascular tissues) form in stems and roots (if present) and as the 'veins' of leaves (if present); they function in the transport of water, ions and organic substances and provide support. The sporophyte is the dominant generation and the gametophyte is reduced. Reduction is greatest in the angiosperms, where the game-tophyte is a few cells attached to the sporophyte or forms a spore (pollen). The dominant group of plants since the Devonian (Figure 2.13).

Class Psilopsida

Only two living genera, both tropical (about 30 species): the leafless *Psilotum* (Figure A.5a) and *Tmesipterus*, which has large leafy shoots.

Very simple plants with sporophytes consisting of dichoto-mously branching stems with sporangia and branching

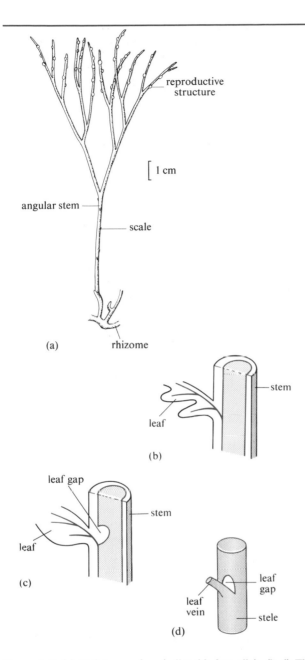

Figure A.5 (a) *Psilotum nudum* (psilopsid; fern ally). (b–d) The origin of leaf veins from vascular tissue in the stem: (b) in Lycopsida and Sphenopsida; (c), (d) in Pteropsida. Vascular tissue is red. In (b) and (c) the stem is cut in half; in (d) the vascular tissue is shown as a solid cylinder.

underground rhizoids. There are no roots. The stem has a central stele and the parenchyma cells under the outer epidermis of *Psilotum* are photosynthetic. The spores germinate into subterranean colourless (i.e. non-photosynthetic) gametophytes (always associated with mycorrhizal fungal hyphae); the male gametes are multiflagellate and swim to the eggs. The earliest land plants known as fossils are assigned to this class (Figure 2.13).

Class Lycopsida (about 1 200 species)

Club-mosses *Lycopodium* (Figure 1.12a and Plate 6a) and *Selaginella*; quill-worts *Isoetes*.

Living species are small plants (but some Palaeozoic fossils were large trees, e.g. *Lepidodendron*); rhizomes with erect branching stems bearing small leaves; true roots present. Sporangia develop on or beside special leaves. *Selaginella* and *Isoetes* produce two types of spores (i.e. they show heterospory): microspores develop into male gametophytes, producing only sperms, and megaspores into female gametophytes, producing only eggs. Male gametes are flagellated and swim to eggs. Gametophytes of club-mosses may be green and grow on the soil surface or may be subterranean and colourless (dependent on mycorrhizal fungi for their nutrition). In some *Selaginella* species the spores germinate while still attached to the sporophyte and the gametophyte plant is very much reduced. *Isoetes* grows in shallow water of non-calcareous ponds and lakes; *Selaginella* and *Lycopodium* species are found in British uplands, but most species are tropical.

Class Sphenopsida (40 species in a single genus, *Equisetum*, horsetails (Figure 1.12b and Plate 6b)

The stems are hollow and grow from rhizomes; the roots grow in whorls from rhizomes and the stems have nodes with whorls of leaves or short branches. Photosynthesis occurs mainly in the stems, which are ridged and have cells with siliceous walls. Sporangia are grouped in cones at the tips of stems. Spores are usually similar, producing small, green, bisexual gametophytes on the surface of soil. The male gametes are multiflagellate and swim to eggs. Sphenopsids were abundant and diverse in the Palaeozoic and some, e.g. *Calamites*, were up to 30 m tall. Living species grow mainly less than 1 m tall and often occur in damp places; they are locally common in Britain. *E. arvensis* (common horsetail) is a garden weed.

Class Pteropsida

The principal feature that unites the ferns, gymnosperms and angiosperms into one class is the leaf development and the direct continuity between the veins of the leaves and the vascular tissue in the stem. The anatomical arrangements of these structures is markedly different from that of the other three classes of tracheophytes (see Figures A.5b, c and d). The subclasses of Pteropsida differ in the structures of the sporophyte and its sporebearing leaves and especially in the form of the gametophytes. Only gymnosperms and angiosperms form seeds.

Subclass Filicidae (about 9 000 species)

Ferns, e.g. *Pteridium* (bracken), *Osmunda* (royal fern), *Asplenium* (spleenwort), *Dryopteris*, *Polypodium*. The stems are usually underground rhizomes but some species have short, erect stems (Figures 1.12c and A.6a); true, simple roots are present. The leaves vary greatly in size, shape and anatomy. Sporangia are borne on leaves (which

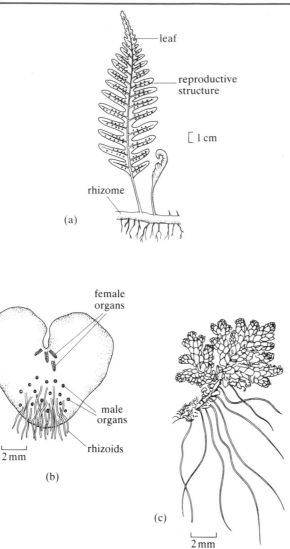

Figure A.6 Filicidae (ferns): (a) *Polypodium vulgare* sporophyte. (b) *Dryopteris* gametophyte (from below). (c) *Azolla* (water fern).

may be structurally different from the ordinary, photosynthetic leaves); spores are usually of one type only and are dispersed by wind. The spores germinate to form a gametophyte (Figures 1.11 and A.6b), often a small, flat, green structure, which may be of several different types, sometimes subterranean. The male gametes are multiflagellated and swim to eggs. The ferns are a widespread and diverse group including large 'tree-ferns' (Plate 6c), delicate 'filmy ferns', and water ferns (e.g. *Azolla*, Figure A.6c).

Subclass Gymnospermidae (about 750 species)

Cycads, gingko, conifers (e.g. *Larix*, *Pinus*) and gnetales. All (except *Welwitschia*) are shrubs or trees and can form secondary xylem (wood); they all produce seeds that are 'naked' (i.e. not contained in an ovary). The spores are of two different types, produced in separate cones. Male cones

produce pollen (microspores) that are dispersed by wind to the female cones. Pollen grains germinate to produce a pollen tube (the male gametophyte) in which male gametes are formed. Female cones produce megaspores which grow into small megagametophytes in which egg cells are formed. After fertilization, zygotes develop into embryos within seeds that can survive adverse conditions and be dispersed by wind or animals (Plate 7a). The outer seed coat forms from cells of the sporophyte and encloses the remains of the megagametophyte, storage tissue and the embryo.

Cycads have unbranched stems and large, palm-like leaves up to 2 m long; they are mainly tropical and grow very slowly. The male and female cones are on separate plants. The male gametes are multiflagellated and large, and swim down the pollen tube to the egg cells.

Gingko: There is only one living species, which is probably now extinct in the wild although there may be a few in south-eastern China. These tall trees have fan-shaped deciduous leaves which are shed annually. The life cycle is basically like that of cycads, with large multiflagellated male gametes, and trees are either male or female, but there are no cones.

Conifers are the largest and most numerous gymnosperms. All are shrubs or trees; some conifer species form vast forests, especially in the northern temperate zone. Californian redwoods are the largest living plants. There is extensive secondary growth of xylem and phloem in the stems, and there is usually a single dominant axis with branches from it. The leaves are typically simple and needle-like or scale-like in form and are usually retained for several years and shed gradually through the year (hence they are mainly 'evergreens'). Many species have thick bark and produce resins, especially if wounded. The pollen is distributed by wind. The male gametes are non-motile. The female cones are larger than male cones and may form on the same or different trees. Pollen germination and seed formation in the female cones may take many months. Seeds are shed when the scales of the female cone separate, sometimes only after heating (e.g. by forest fires).

The living genera of gnetales (*Ephedra*, *Welwitschia* and *Gnetum*) have many features in common with angiosperms.

Subclass Angiospermidae (about 285 000 species)

The pollen and ova of flowering plants form in flowers (Figure 1.13), not cones; the pollen is produced in the anthers of stamens and the ova are formed inside ovules that are inside an ovary. Pollen is dispersed by wind or by animals (usually insects) and germinates on a stigma to form pollen tubes (microgametophytes) that grow down the style into the ovary. The male gametes are non-motile nuclei; the egg is a nucleus produced by division within an embryo sac (the megagametophyte). The seed (Figure 1.14) consists of the embryo surrounded by endosperm or other nutrient-storage tissue and a seed coat produced by the parent sporophyte; the ovary and other parts of the flower form a fruit around the seed. Angiosperm vascular tissue comprises xylem vessels and sieve tubes (instead of the xylem tracheids and sieve cells in gymnosperms and other lower tracheophytes) and the leaves are of many different forms.

Superorder Monocotyledonae (about 48 500 species) The principal veins in the leaves lie parallel to each other and there is one cotyledon (a nutrient-storing leaf-like structure) in the embryo. The vascular bundles in the stem are scattered irregularly and there is no true secondary thickening (formation of rings of wood in the stem). There are numerous fibrous roots of similar size. The parts of the flowers are arranged in threes or multiples of threes. Monocotyledon families include: Liliaceae (lilies); Orchidaceae (orchids, Plate 7e); Palmaceae (palms); Cyperaceae (sedges, rushes and papyrus reeds (Plate 15f)); Gramineae (grasses and reeds (Figure A.7)). The growth zones (meristems) are at the bases of the leaves of grasses so they continue to grow when cropped by grazing animals. Grass flowers are usually wind-pollinated and the seeds contain stores of starch and protein. The major cereal food crops such as wheat, maize and rice are grasses.

Figure A.7 Angiospermidae: Monocotyledonae, Graminaceae. *Festuca ovina*, a grass.

Superorder Dicotyledonae (about 236 500 species) The principal veins of the leaves branch from the midrib or from the base, forming a network and there are two seed leaves (cotyledons) in the embryo. The stem has vascular bundles in a single cylinder and the cambium divides to form secondary thickening. In perennials, the cambium adds a new ring of wood in each growing season. The root system consists of a large primary root (which may become a tap root storage organ) with branch roots growing from it. The parts of the flowers are arranged in twos, fours or (usually) fives. There are diverse growth forms, from tiny 'ephemer-als' to large forest trees. The majority of living plants are dicotyledons, including the families: Ranunculaceae (e.g. buttercups); Rosaceae (e.g. rose, apple, blackberry (Plate 7c)); Leguminosae (e.g. pea, bean, laburnum); Labiatae (e.g. deadnettle (Plate 7d), mint); Umbelliferae (e.g. carrot, celery, hogweed); Cruciferae (e.g. cabbage, mustard); Compositae (e.g. daisy, knapweed, sunflower).

KINGDOM FUNGI

Eukaryotic heterotrophic organisms (i.e. with membrane-bound nuclei and mitochondria) that never contain chlorophyll and never photosynthesize (Figure 1.15). Most are saprophytes that live on dead organic matter (Plate 8), or are parasites; some form mutualistic relationships with algae (lichens) or with flowering plants (mycorrhizae).

Division MYXOMYCOTA

Myxomycetes (slime moulds) (Figure A.8) resemble protistans and are sometimes included in the class Sarcodina. No cell walls are present except in spores (which have cellulose walls). Small amoeboid feeding forms (each with one nucleus) engulf bacteria and rotting wood. Sometimes they aggregate into a multinucleate plasmodium (sometimes with half a million nuclei) that eventually grows one or more fruiting bodies (sporangia) in which small uninucleate 'spores' with cellulose walls are formed by meiosis. The sporangium bursts and releases the spores which become amoeboid, often with flagella, and fuse in pairs to form

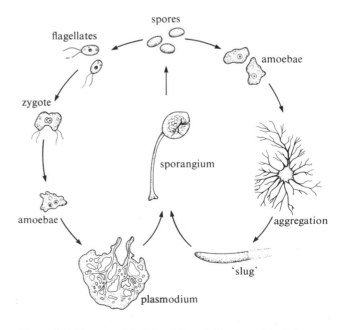

Figure A.8 Two possible life cycles of slime moulds (Myxomycota).

153

zygotes which become the small feeding amoebae (see Figure A.8). All slime moulds (e.g. *Physarum*, *Dictyostelium*) are terrestrial, living in damp places (e.g. on rotting logs); *Plasmodiophora* is parasitic, causing 'club root' of *Brassica* plants.

Division EUMYCOTA (about 100 000 species)

True fungi have rigid cell walls containing chitin or (rarely) cellulose; the cell contents stream actively and there are never any large vacuoles. Some true fungi are unicellular but usually cells (diameter 3 to 5 μm) are joined end-to-end to form filaments (hyphae) that interwine as a mycelium. Cells secrete digestive enzymes and absorb the products of digestion. Fungi are decomposers of organic material, especially in terrestrial and freshwater environments; some species are economically important parasites, mainly of plants. There are three classes.

Class Phycomycetes (about 50 000 species)

The hyphae lack cross walls. The major groups are:

Subclass Chytridiomycetes

Aquatic, usually unicellular fungi, with chitinous cell walls. The gametes and spores are motile with flagella. Many are parasites of planktonic algae. *Rhizophydium* (Figure A.9a).

Subclass Oomycetes

Hyphae are multinucleate, with cellulose walls (Figure A.9b). The gametes are non-motile and the spores have two flagella. Oomycetes are decomposers in aquatic and terrestrial habitats, e.g. *Saprolegnia* (aquatic fungus), and parasites, e.g. *Phytophthora infestans* (potato blight).

Subclass Zygomycetes

Hyphae are multinucleate, with chitinous walls. Flagella are absent. Spores are formed either in a capsule (sporangium) or naked in chains (Figures 1.15a and b and A.9c, d and e). *Mucor* (bread mould), *Entomophthora* (insect parasite), *Pilobolus* (soil and dung fungus).

Class Ascomycetes (about 30 000 species)

Unicellular yeasts, e.g. *Saccharomyces* (Figure A.10) or mycelial forms with chitinous cell walls; the hyphae are haploid or form binucleate dikaryons. Dikaryote hyphae may aggregate to produce fruiting bodies that contain reproductive structures, called asci, in which, after fusion of nuclei and meiosis, eight haploid ascospores are formed; some fruiting bodies are quite large (up to 10 cm). All species produce asexual spores that are often coloured. Many species form lichens (Figure 2.21). Examples of mycelial forms are: *Penicillium* (with green spores; Figures A.11a and 1.15c), *Aspergillus* (Figure A.11b), *Neurospora* (bread mould with red or black spores), *Claviceps* (ergot), *Ascobolus* (cup fungus; Figure A.11c), *Morchella* (morel).

Class Basidiomycetes (about 13 000 species)

The hyphae have chitinous walls and cross-walls with pores; in dikaryons the cross-walls form clamp connections (Figure A.12). The dikaryon forms fruiting bodies, sometimes large and conspicuous (e.g. mushroom *Agaricus*), in which basidiospores are formed after fusion and meiosis. Some are parasitic (e.g. *Puccinia* (rust) and smuts (Figure A.13)) and produce several types of spores; some are serious pathogens of crop plants. Others are saprophytes (Figure 1.15d and Plate 8b), able to digest complex organic substances including lignin; many form mycorrhizal associations with trees. *Ganoderma* (bracket fungus), *Lycoperdon* (puff balls), *Panaeolus* (Figure A.14a), *Boletus* (edible ceps; Figure A.14b), *Amanita* (fly agaric).

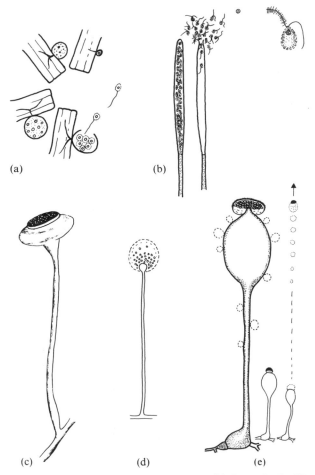

(a)   (b)

(c)   (d)   (e)

Figure A.9 The sporangia of Phycomycetes. (a) Stages in the life cycle of *Rhizophydium* (chytridiomycete). (b) *Saprolegnia* (oomycete) releasing spores. (c) *Pilaira*. (d) *Mucor*. (e) *Pilobolus*. (c), (d) and (e) are zygomycetes; (d) is a bread mould; (c) and (e) grow on cattle dung; (a) and (b) are freshwater fungi.

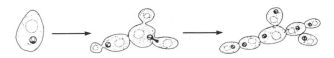

Figure A.10 Ascomycetes: *Saccharomyces* (yeast).

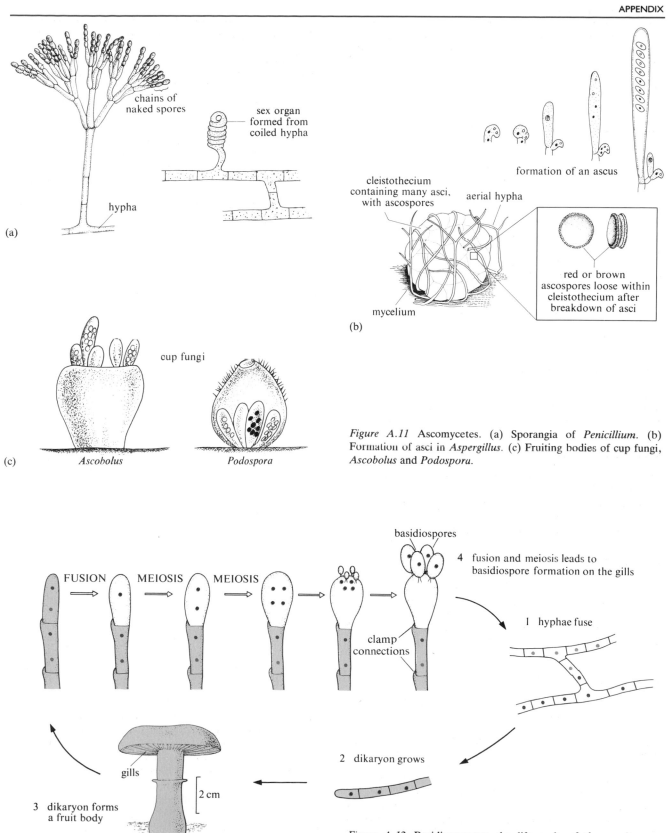

(a)

chains of naked spores

hypha

sex organ formed from coiled hypha

formation of an ascus

cleistothecium containing many asci, with ascospores

aerial hypha

mycelium

red or brown ascospores loose within cleistothecium after breakdown of asci

(b)

(c)

cup fungi

Ascobolus

Podospora

Figure A.11 Ascomycetes. (a) Sporangia of *Penicillium*. (b) Formation of asci in *Aspergillus*. (c) Fruiting bodies of cup fungi, *Ascobolus* and *Podospora*.

FUSION MEIOSIS MEIOSIS

basidiospores

4 fusion and meiosis leads to basidiospore formation on the gills

clamp connections

1 hyphae fuse

2 dikaryon grows

gills

2 cm

3 dikaryon forms a fruit body

Figure A.12 Basidiomycetes: the life cycle of the mushroom, *Agaricus*. The dikaryon is shown in pink.

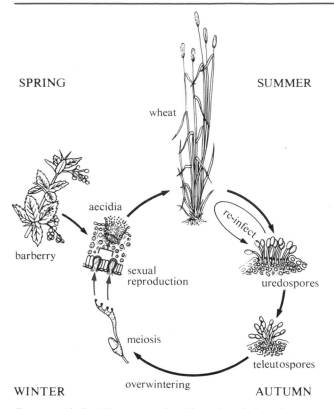

SPRING

SUMMER

wheat

aecidia

barberry

re-infect

sexual
reproduction

uredospores

meiosis

teleutospores

overwintering

WINTER

AUTUMN

Figure A.13 Basidiomycetes; the life cycle of the wheat rust fungus, *Puccinia graminis*. The definitive host is wheat, *Triticum vulgare*, and the secondary host is barberry, *Berberis vulgaris*, a hedgerow or garden shrub. Rusts spread rapidly in summer among wheat plants by means of wind-blown uredospores. They overwinter on wheat straw as thick-walled, black teleutospores, which germinate in spring after fusion of nuclei followed by meiosis and production of sporidia (basidiospores). The sporidia are dispersed by wind and germinate only on barberry leaves, producing a mycelium and then structures called spermogonia where sexual reproduction occurs by fusion of 'male' spores with the tips of 'female' hyphae. Binucleate cells are produced which divide to form aecidia (cluster cups) on the undersides of leaves and from which chains of aecidiospores are released that infect wheat plants.

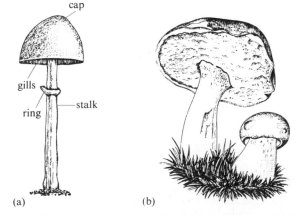

cap

gills

ring

stalk

(a)

(b)

Figure A.14 Fruiting bodies of Basidiomycetes: (a) *Panaeolus*. (b) *Boletus edulis*.

KINGDOM PROTISTA (sometimes called Protozoa)
(at least 40 000 species)

Heterotrophic, unicellular eukaryotic organisms. There is overlap with the other eukaryote kingdoms and confusion about nomenclature because botanists and zoologists use different names for the same organisms. There are several classification schemes; here four classes are recognized.

Class Mastigophora

Single-celled organisms, typically with flagella, that never produce spores. Some have chloroplasts containing photosynthetic pigments and may also be included in the kingdom Plantae (with different names and status for the taxa).

Subclass Phytomastigina

Photosynthetic species that resemble unicellular algae in form and habits and some colourless, 'animal-like' forms.

Order Euglenoidida Photosynthetic green forms, e.g. *Euglena* (Figure 1.7a) and some non-photosynthetic colourless forms, e.g. *Peranema* (Figure 1.8a). See kingdom Plantae, division Euglenophyta.

Order Phytomonadida A large group including green flagellates, e.g. *Chlamydomonas* (Figure 1.7b), *Volvox* (Plate 3d) and some colourless forms, e.g. *Polytoma* (Figure 1.8b). See kingdom Plantae, division Chlorophyta, class Chlorophyceae.

Order Dinoflagellida Mostly photosynthetic but there are colourless heterotrophic and parasitic species. Common in freshwater and marine plankton, e.g. *Ceratium*, *Gymnodinium*, *Gonyaulax*, *Noctiluca* (Figures 1.6a and b). Some are symbionts called zooxanthellae (e.g. in corals). See kingdom Plantae, division Pyrrophyta.

Subclass Zoomastigina

Non-photosynthetic protistans (Figure A.15) with more complex flagella and associated structures than in Phytomastigina.

Order Protomonadida Small protistans, usually with one or two flagella; free-living, e.g. choanoflagellates, or parasitic forms, e.g. *Trypanosoma* (Figure A.15a) which causes sleeping sickness.

Orders Trichomonadida and Diplomonadida Usually parasites in the guts of vertebrate and invertebrate animals; with four or more flagella arranged in various ways, e.g. *Giardia* (Figure A.15b).

Order Hypermastigida With 50 or more flagella often arranged in a complicated pattern; one nucleus only. They live as symbionts in the guts of cockroaches (insect order Orthoptera) or termites (insect order Isoptera), e.g. *Trichonympha* (Figure A.15c) which can digest wood fragments.

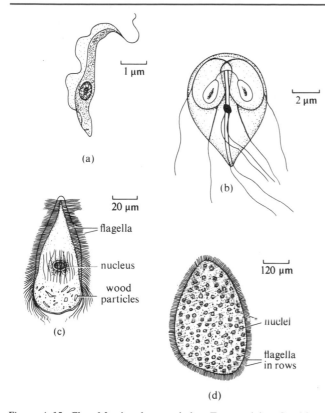

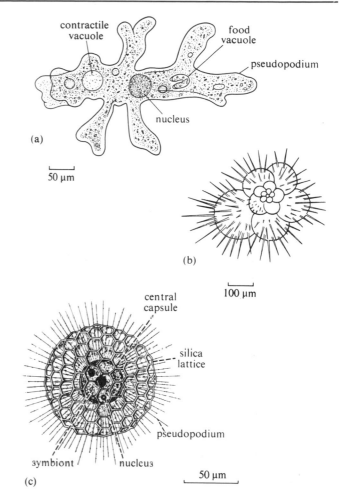

Figure A.15 Class Mastigophora, subclass Zoomastigina. Symbiotic protistans living in vertebrates and insects: (a) *Trypanosoma* (Protomonadida), a blood parasite which causes sleeping sickness in humans. (b) *Giardia* (Diplomonadida) which lives in human guts. (c) *Trichonympha* (Hypermastigida), which lives in the guts of termites and aids digestion of wood fragments. (d) *Opalina* (Opalinida), a very large protistan found in the rectum of frogs.

Order Opalinida With very numerous flagella (sometimes called cilia) in rows over a flattened body; they have many similar nuclei. Found in guts of frogs and toads, e.g. *Opalina* (Figure A.15d). This order is sometimes included in the class Ciliata.

Class Sarcodina

Protistans that usually feed and move by means of pseudopodia and have relatively few organelles; some have flagellated stages in the life cycle and they never produce spores. The classification is based on the types of pseudopodia and skeletons (if present) and presence or absence of flagella at some stage (Figure A.16).

Order Amoebida Naked cells producing broad, blunt pseudopodia; mainly freshwater forms, e.g. *Amoeba* (Figure A.16a), or gut parasites, e.g. *Entamoeba*.

Order Foraminiferida Cells with calcareous shells, usually of several chambers, with pores through which a net of fine anastomosing pseudopodia projects. There is often

Figure A.16 Free-living, aquatic sarcodine protistans: (a) *Amoeba* (Amoebida), which lives in still, shallow freshwaters. (b) *Globigerina* (Foraminiferida), which lives in marine plankton. (c) *Heliosphaera* (Radiolarida), which lives in marine plankton.

alternation of sexual and asexual generations, which may look different; the gametes have flagella. Mostly marine, in the plankton or in deeper water. *Polystomella, Globigerina* (Figure A.16b).

Order Radiolarida The skeleton is made of silica; the body is divided into central capsule and outer vacuolated cytoplasm, with fine, branching and anastomosing pseudopodia. They may be multinucleate and may have symbiotic zooxanthellae (dinoflagellates) in the outer cytoplasm. All are marine and planktonic. *Heliosphaera* (Figure 1.5c and Figure A.16c).

Order Heliozoida The cells may be naked or with a gelatinous coat or skeleton of silica; stiff contractile pseudopodia project in all directions. The body is usually divided into inner denser cytoplasm and outer vacuolated cytoplasm; symbiotic zoochlorellae (green algae) are often present. Typically freshwater forms. *Actinosphaerium* (sun animalcules).

Class Ciliata (about 6 000 species)

With cilia or ciliary organelles at some stage of the life cycle; typically with two types of nuclei, micronuclei (which divide by mitosis and by meiosis during sexual reproduction) and meganuclei (essential for the survival of individuals but do not divide by meiosis). Two or more subclasses recognized mainly on the basis of the arrangement of the cilia (Figure A.17), especially 'membranellae' (cilia that adhere in groups and form paddles).

Subclass Holotricha

Body of various shapes, usually with a small buccal cavity (where food enters the body) and uniform ciliation. The orders include:

Order Hymenostomatida   With ventral buccal opening, e.g. *Paramecium* (Figure 1.5a and Plate 2a).

Order Suctorida   Sessile as adults and without cilia; feed by suctorial tentacles, e.g. *Discophrya*.

Order Peritrichida   Usually attached to substratum by a stalk or basal disc, with cilia reduced or absent over the body; but conspicuous oral cilia forming a counterclockwise buccal spiral, e.g. *Vorticella* (Plate 2b and Figure 1.5b).

Subclass Spirotricha

With highly developed membranellae that form clockwise spirals at the oral end (Figure A.17); cilia elsewhere on the body are usually reduced or specialized (e.g. to form 'cirri'). The orders include:

Order Heterotrichida   Body covered with short cilia and of diverse form. Some large (up to 3 mm long), e.g. *Stentor* (Figure A.17a).

Order Tintinnida (about 900 species)   Shell secreted around the cell, which has membranellae and other specialized cilia. Mainly in marine plankton, e.g. *Tintinnopsis*.

Order Hypotrichida   The cilia are dorsal 'bristles' and cirri on the ventral surface; they are usually benthic, 'walking' on the cirri. *Amphisia*, *Euplotes* (Figure A.17b).

Order Entodiniomorphida   Mostly obligate anaerobes, able to live by fermentation and to digest cellulose and many are gut symbionts of mammals, especially herbivores; they have the most complex internal structure of all protistans. *Epidinium* (Figure A.17c).

Class Sporozoa

Parasitic protistans without flagella or cilia (except occasionally on spores); often with complex life cycles, e.g. the malarial parasites, *Plasmodium*, and the organisms that cause coccidiosis, *Eimeria*.

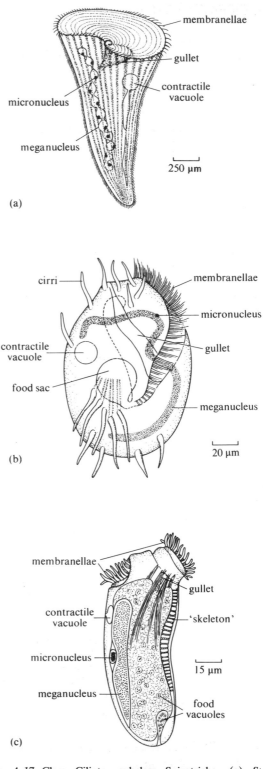

Figure A.17 Class Ciliata, subclass Spirotricha. (a) *Stentor* (Heterotrichida), which lives in still, shallow freshwaters. (b) *Euplotes* (Hypotrichida), which lives in still, shallow waters. (c) *Epidinium* (Entodiniomorphida), a symbiotic protistan which lives in the guts of mammals.

KINGDOM ANIMALIA

Multicellular (metazoan) eukaryotic organisms, most of which feed on or are parasites of, other living organisms. The principal phyla are listed in Table A.1.

Table A.1 Some phyla of the kingdom Animalia

Phylum	Estimated number of living species	Common names
Porifera (Parazoa)	>4 000	sponges
Cnidaria (Coelenterata)	11 000	jellyfish, sea-anemones, corals, hydroids
Ctenophora	80	comb-jellies, sea-gooseberries
Platyhelminthes	15 000	flatworms, flukes, tapeworms
Nemertina (Rhynchocoela)	600	proboscis worms, ribbon-worms
*Acanthocephala	800	spiny-headed worms
*Entoprocta	60	
Nematoda	80 000	roundworms
Mollusca	110 000	snails, mussels, squids, octopus
Annelida	8 800	segmented worms
Pogonophora	45	
Arthropoda	>1 000 000	insects, crustaceans, spiders
Onychophora	70	
*Tardigrada	170	water-bears
*Linguatulida	60	tongue-worms
*Echiuroidea	80	
*Ectoprocta	4 000	moss-animals
*Priapulida	8	
*Phoronida	15	
*Brachiopoda	310	lamp-shells
*Sipunculoidea	275	
*Chaetognatha	60	arrow-worms
Echinodermata	6 000	starfish, brittle-stars, sea-urchins, sea-cucumbers, feather stars
Hemichordata	100	acorn worms
Chordata	54 000	lancelets and sea-squirts (2 100 species) and vertebrates (fish, amphibians, reptiles, birds and mammals)

*Not covered in this Survey or in the text.

Phylum PORIFERA (sometimes called sub-kingdom Parazoa; more than 4 000 species)

Very simple multicellular animals (sponges), sometimes considered to be colonies of cells, mainly marine (Plate 9), a few in freshwaters. Water enters by smaller pores (ostia) and leaves by larger pores (oscula) by currents created and filtered by choanocytes on the inner body wall (Figure 1.17). An outer layer of cells covers a matrix containing amoebocytes that secrete spicules of silica or calcium carbonate.

Phylum CNIDARIA (sometimes called Coelenterata; about 11 000 species)

The radially symmetrical body is composed of two layers of cells (an outer ectoderm and inner endoderm), separated by a gelatinous mesogloea (Figures A.18 and 1.18). The gut (enteron) may be simple or divided into pouches and canals and there is usually a single opening surrounded by tentacles. Undigested remains are discharged through the mouth. Unique kinds of cells include musculoepithelial cells and cnidoblasts, stinging cells mostly in the ectoderm, some of which inject toxic substances that can paralyse. All cnidarians are aquatic, most are marine carnivores, feeding on animals that blunder into the tentacles and are stung and paralysed. Some are filter feeders. There are two basic shapes: polyps that are like sacs and normally live attached to rocks or other objects; umbrella-shaped medusae have much thicker mesogloea and are usually much larger than polyps and are free-swimming.

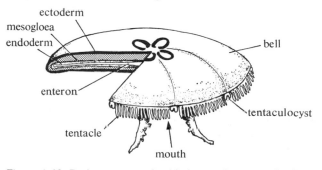

Figure A.18 Basic anatomy of cnidarian medusa, *Aurelia* (jellyfish). The mesogloea is thick and elastic and the swimming muscles are ectodermal and on the oral surface. Tentaculocysts are small sense organs that detect tilting. The mouth leads into an enteron that consists of canals and four gastric pouches containing digestive filaments. There are four horseshoe-shaped gonads visible through the transparent bell.

Class Hydrozoa

Typically have both polyp and medusa stages in life cycle but many have polyps only, and some have medusae only. Some form large elaborate colonies, e.g. *Physalia* (Portuguese man o' war). Fire corals (or stinging corals) are colonial hydrozoans that secrete a white calcareous external skeleton. *Hydra* polyps (Figure 1.18a) are very small and lack mesenteries.

Class Scyphozoa

Large jellyfish with massive mesogloea; typically the larvae develop into medusae but some have stages in the life cycle comparable with polyps, e.g. *Aurelia* (Figures 1.18b and c and A.18).

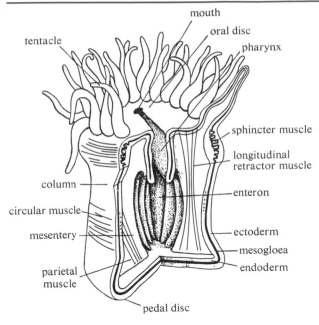

Figure A.19 Basic anatomy of a cnidarian polyp (sea-anemone). The enteron cavity is partly divided by several vertical partitions called mesenteries, consisting of mesogloea with endoderm on both sides, and containing the retractor muscles. The pedal disc attaches the anemone to the substrate.

Class Anthozoa

Polyps only; solitary sea anemones (Figure A.19), e.g. *Metridium* and reef-building stony corals (Scleractinia or Madreporaria) (Figure A.20). Corals are small anemones in which the ectoderm secretes a calcareous skeleton outside the animal. They usually live in colonies fixed to the substratum and the skeletons are usually white and massive. The living polyps usually have many tentacles with cnidoblasts and are brilliantly coloured, due to pigments in the cells. The polyps reproduce sexually by forming gametes and asexually by budding. The shape of the colony is related to the type of budding and the degree of separation of the resultant polyps; extremes are the 'brain corals' (*Meandrina*; Figure A.20b) where the polyps do not separate and rows of mouths are surrounded by rows of tentacles. A few stony corals are solitary, e.g. *Fungia*, which is stalked when young but becomes free-living when mature.

Phylum CTENOPHORA

Radially symmetrical animals with thick mesogloea and eight rows of ciliary plates (comb plates). Cnidoblasts absent. All marine free-swimming or planktonic predators, that never form colonies. *Pleurobrachia* (sea-gooseberry).

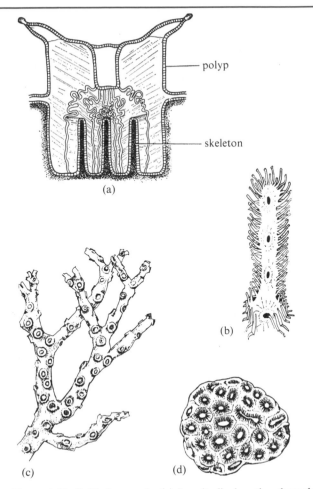

Figure A.20 Cnidarian corals. (a) Longitudinal section through a single coral polyp and its skeleton; (b) A row of four fused polyps of the reef-building brain coral *Meandrina* with the tentacles extended. A large colony consists of hundreds of intersecting rows of polyps that resemble a human brain. (c) The skeleton of *Oculina*, the ivory bush coral. In life, the entire structure would be covered with living tissue with tentacles and a small mouth at each round polyp. (d) The skeleton of the reef-building coral *Favia*. Like those of *Meandrina*, the polyps form side by side and grow radially, forming a clump, but in contrast to *Meandrina*, each *Favia* polyp has a complete ring of tentacles.

Phylum PLATYHELMINTHES

The body consists of three layers of cells: ectoderm covers the outside of the body; endoderm forms the lining of the gut; mesoderm lies between ectoderm and endoderm. Mesoderm forms muscles, reproductive organs, and a general packing of cells called parenchyma. There is no blood system and no cavity between gut and body wall except for the spaces inside the reproductive organs and their ducts. The gut has only one opening. Platyhelminths are usually hermaphrodite and lay yolky eggs, and there are several larval stages. Most are free-living carnivores or parasites.

Class Turbellaria

Small, typically free-living flatworms, called planarians, that usually feed by seizing animals with an eversible or a protrusible pharynx (Figure 1.19). Ectoderm is normally ciliated. This class is further subdivided, primarily on the basis of the branching of the gut: rhabdocoels have a single branch; polyclads (Figure A.21b) have a gut with more than three branches and are all marine; triclads (Figure A.21a) have a three-branched gut and live in the sea, in freshwaters, and a few in damp places on land, e.g. *Dendrocoelum*.

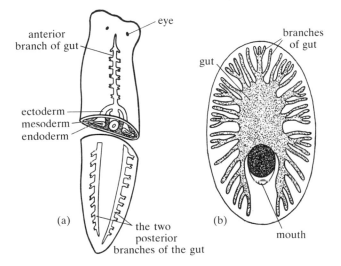

Figure A.21 Free-living platyhelminths. (a) A triclad flatworm, showing gut and layers of tissues in the body. (b) A polyclad flatworm showing the gut.

Class Trematoda

Similar in shape to flatworms but one or two suckers present and sometimes an extra attachment organ bearing hooks. Gut with two branches. Ectoderm is nonciliated, usually bearing minute spines. Flukes are divided into two subclasses: Monogenea and Digenea. The former have a simple life cycle with one vertebrate host and a free-swimming larva while the latter have at least two hosts, one a vertebrate and the other normally a mollusc that harbours the larval stages, e.g. *Schistosoma* (Figure 2.22 and Plates 10b–f); *Clonorchis* (Chinese liver fluke; Figure A.22a).

Class Cestoda

The scolex (head) bears suckers and/or hooks, and a body of many units (proglottids), each of which bears a set of reproductive organs that can become detached when mature, leading to dispersal of eggs. Tapeworms lack a gut completely; they absorb food in solution through the ectoderm, which is living and non-ciliated but bears microvillus-like structures. All are parasites; adults usually live in the gut of vertebrates, usually with at least one larval form also living in a vertebrate but some species have larval forms in invertebrates, particularly crustaceans and molluscs. *Taenia* (human tapeworm; Figure A.22b–d) has larval stages in viscera and muscles of pigs and the adults live in the human intestine.

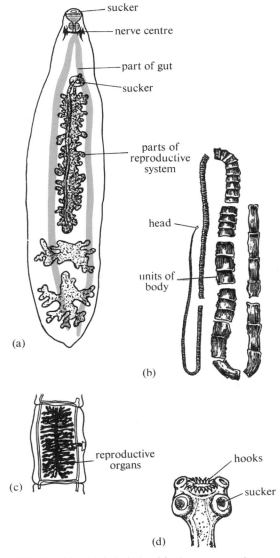

Figure A.22 Parasitic platyhelminths. (a) The anatomy of a trematode fluke, *Clonorchis*. (b–d) Cestode tapeworm *Taenia* (b) parts of the whole worm; (c) one ripe proglottis; (d) the scolex (head).

Phylum NEMERTINA (also called Rhynchocoela; about 600 species)

A homogeneous group called proboscis worms or ribbon worms (Figure A.23) with a round or flattened body composed of three layers of cells: ciliated ectoderm, endoderm and mesoderm. There is a thick basement membrane between ectoderm and mesoderm, forming an elastic but firm cover to the muscles, which form several layers. Parenchyma is present between muscles, reproductive and excretory

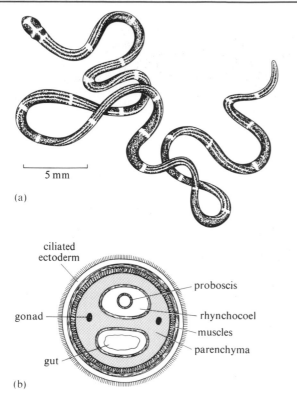

5 mm

(a)

ciliated
ectoderm

proboscis

gonad

rhynchocoel

muscles

parenchyma

gut

(b)

Figure A.23 Nemertina, *Tubulanus*: (a) whole worm; (b) diagrammatic transverse section near the anterior end of body.

organs. The gut has two openings: mouth and anus. The proboscis, lying in a special cavity, the rhynchocoel, above the gut, can be everted and retracted by antagonistic muscles with the rhynchocoel acting as a hydroskeleton; it is coiled round prey which is then drawn back towards the mouth. The blood system is closed, with contractile vessels and sometimes with the respiratory pigment haemoglobin in solution in the blood. Both sexes have many paired gonads which are simple sacs opening by pores to the exterior and shedding gametes into the sea where fertilization occurs. There is a planktonic ciliated larva, called the pilidium. The size range is from a few mm to 30 m. Nemertines live mainly on the floor of shallow seas of the temperate zone; a few live in freshwaters or on land.

Phylum NEMATODA (about 80 000 species)

Roundworms (Figures 1.20 and A.24a) are round in section and composed of three layers of cells: ectoderm, endoderm and mesoderm (which forms muscles, reproductive organs and excretory organs). There is a cavity, the pseudocoel, between the outside of the gut and the body muscles, which is not lined by epithelium nor divided into compartments. The gut has two openings, mouth and anus, but the walls are muscular only in the pharynx. There is no blood system. The body is wholly or partly covered by cuticle and cilia are absent. Roundworms are long and slender, moving by whip-like lashings of the body. Only longitudinal body

muscles are present, with contractile fibres in the outer part of the muscle cell and a nucleus in the inner part. The excretory system consists of two lateral longitudinal canals opening by a single pore near the anterior end. The nervous system consists of a ring round the pharynx and longitudinal cords passing down the body. Sexes are separate; the gonads are tubular and sometimes have two branches. Fertilization is internal and the eggs typically begin to develop in the oviduct and may hatch into the first larval stage within the female (i.e. many species are ovoviviparous). There are three or four larval stages. The numbers of nuclei in adults are remarkably constant for each species. Roundworms are widely distributed and often abundant in soil, freshwater, saltwater and as parasites of plants and animals; free-living and parasitic forms are similar in basic structure.

Ascaris (Figures A.24a and b) is a parasite of pigs and humans; adults live in the intestine. The eggs are shed in the faeces and develop in the soil where they can survive for many months. Infection is by ingestion of eggs containing embryos, which hatch in the intestine and burrow into its wall. The larvae enter the lymphatic system and eventually reach the blood system and pass through the heart to the lungs; they break into the lung alveoli and eventually (about 10 days after infection) move up the trachea and down the oesophagus, through the stomach to the intestine where they settle down to become adults. The passage of large numbers of larvae through the lungs may result in pneumonia; adults in the intestine produce toxic secretions so there may be nervous symptoms in addition to obstruction of the gut.

Ancylostoma is one of the hookworms parasitic in humans. Adults attach themselves to the walls of the intestine (Figure A.24c) and suck blood continuously; hosts may become anaemic. Eggs are shed with the host's faeces and hatch in moist soil; third stage larvae usually enter human hosts by boring through the skin of the leg and foot. They enter veins and pass through the blood system to the heart and then to the lungs, breaking into the lung alveoli and, like *Ascaris* larvae, moving up the trachea and down the oesophagus, through the stomach to the intestine. Other hookworms infect domestic animals, including horses, sheep and fowl.

Adult *Wuchereria* are internal parasites of humans. Eggs develop inside the adults and first-stage larvae (microfilariae) live by day in the lymphatic or blood vessels of the lungs and other viscera but by night in vessels in the skin. Mosquitoes biting at night suck blood containing larvae which move through the insect's stomach and into its flight muscles, where they grow. The third-stage larvae move to the mosquito's mouthparts and through the wound it makes into the blood of the next human victim; they move from the blood into the lymphatic system and mature. *Wuchereria* and related worms produce diseases called filariasis and elephantiasis (swelling, usually of a leg, as a result of blocking of lymphatic ducts).

Trichinella is a rat and pig parasite that sometimes infects humans who eat undercooked pork. Adults live in the small intestine, but ripe (ovoviviparous) females move into the lymphatic spaces of the intestinal wall, where they may each produce up to 10 000 larvae which move into the blood system and are distributed to the muscles, where they

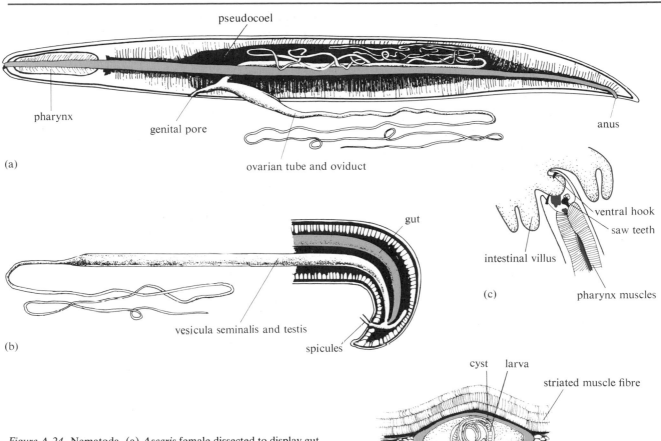

Figure A.24 Nematoda. (a) *Ascaris* female dissected to display gut and gonads. (b) Posterior end of male *Ascaris*. (c) *Ancylostoma*: longitudinal section through head attached to wall of intestine. (d) *Trichinella*: encysted larva in pig muscle tissue.

encyst, leading to degeneration of muscles (see Figure A.24d). The next host is infected by eating the muscles of the previous host; the cysts dissolve in the digestive juices and the larvae soon become mature in the new host's intestine. Infected pork has been found with 15 000 encysted *Trichinella* per gram of muscle.

Phylum MOLLUSCA (about 110 000 species)

The unsegmented body is divided into a head-foot and a visceral mass (Figure A.25). The foot is usually very muscular and is often the organ of locomotion. The visceral mass is covered with a 'mantle' which usually extends as a fold or folds and encloses a mantle cavity, which contains two gills (called ctenidia). The external layer of the mantle often secretes a shell which may consist of one or two valves, sometimes of several plates. The coelom is restricted to the cavities of the reproductive organs, kidneys and pericardium. The blood system is open and the main body cavity is a haemocoel. The 'head' bears the mouth, often containing a rasping radula, and usually sense organs. The gut has a mouth and an anus and a large digestive gland. The main classes are:

Class Gastropoda

Shell usually of one piece (Figure 1.26 and Plates 13b and c) or absent (Plate 13d); foot flat, movement by creeping; head distinct, with tentacles, a radula and often stalked eyes; visceral mass often coiled in coiled shell (Figure 1.26). Some are hermaphrodite. Most marine forms have a veliger larva (Figure 2.6). The class includes a great diversity of forms, mostly aquatic or terrestrial animals confined to damp places, e.g. snails, slugs, whelks, limpets (Figure A.26).

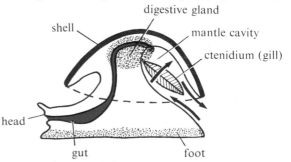

Figure A.25 Mollusca. The basic structure of a hypothetical ancestral mollusc. The arrows show the direction of flow of water through the mantle cavity and over the gill.

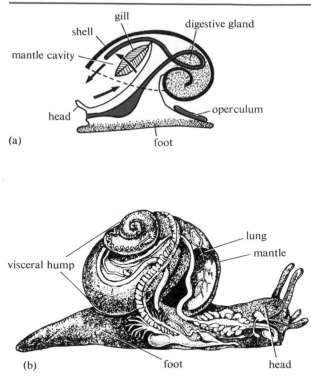

(a)

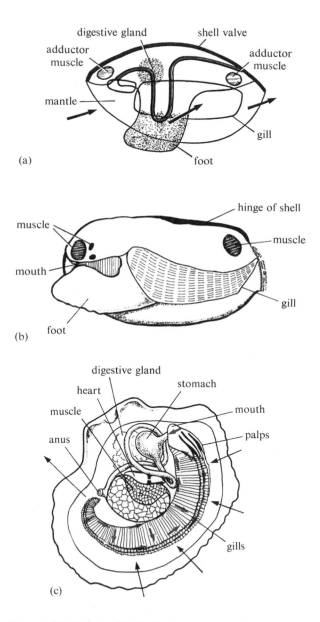

Class Bivalvia (also called Lamellibranchiata or Pelecypoda)

Shell of two valves that can completely enclose the body. Head reduced and sense organs and radula absent from it. Gills very much enlarged with ciliary tracts that function in food collection as well as respiration. Foot adapted for burrowing, or reduced (Figures 1.27 and A.27a). Large adductor muscles close shell valves. Sexes usually separate. Most are marine but some species live in freshwaters. The

(b)

Figure A.26 Gastropoda. (a) The basic anatomy of a gastropod mollusc. (b) The common snail *Helix* (a pulmonate), dissected to show the major organs.

Order Prosobranchiata
Gills in the mantle cavity and in front of the heart; visceral hump coiled; shell usually present and may be closed by an operculum; sexes are usually separate. The largest order, mostly marine. Whelks, limpets, periwinkles, cowries (Plate 13c).

Order Opisthobranchiata
All marine, with single gill, if present, posterior to heart. Mostly hermaphrodite carnivorous forms living near seashore; some are pelagic. Shell reduced or absent. *Aplysia* (sea-hare), *Tritonia* (sea-slug).

Order Pulmonata
Freshwater and terrestrial forms without gills; mantle cavity functions as lung. With or without a shell; operculum absent. Mostly herbivorous and hermaphrodite. Snails (Figure A.26b; Plate 13b), slugs (Plate 13d).

Class Polyplacophora

Internal anatomy similar to that of prosobranch gastropods but the shell consists of several articulating plates and the gills are serially arranged. Not a very diverse group, but some species are abundant. Chitons (Plate 13a).

Figure A.27 Bivalvia. (a) The basic anatomy of a bivalve mollusc. (b) The swan mussel *Anodonta* with the left shell and left mantle removed to show the gills and foot. (c) The oyster *Ostrea* with the upper shell and mantle removed, dissected to show the viscera. Red arrows show the direction of the ciliary currents; black arrows show the direction of water flow.

majority feed by filtering plankton from a stream of water passed through the gills (ctenidia) by ciliary action. The water current enters between the folds of the mantle round the gape of the shell or through a restricted channel, the inhalant siphon, then passes through the gills and leaves through an exhalant siphon which may be short or long. The two siphons may be separate (as in the cockle, *Cardium*) or bound together (as in the clam, *Mya*, and the razor shell, *Ensis*). The freshwater swan mussel, *Anodonta* (Figure A.27b) has sieve-like gills and no specialized siphons. Figure A.27c shows the ciliary currents over the gills of the oyster, *Ostrea*, which also lacks an inhalant siphon but its gills are less solid. The further classification of bivalves depends on gill structure, on the arrangement of the shell adductor muscles (e.g. the differences between *Ostrea* and *Anodonta*), and on the shape of the hinge between the two shell valves.

Class Cephalopoda

Head with radula and beak, and usually with large eyes and well-developed brain. Foot forms sucker-bearing tentacles round the mouth and the funnel. Haemocoel is greatly reduced so there is no significant body cavity and the blood system includes capillaries. Shell is large and chambered in *Nautilus* and in many extinct forms (ammonites etc.) and reduced and internal or absent in most living forms. Mantle cavity contains one or two pairs of gills. Mantle is thick and muscular and can expel water through the funnel. Many species have elaborate chromatophores and change colour rapidly, some secrete ink when disturbed. All are marine, usually swimming or bottom-living carnivores, e.g. octopus, cuttlefish, squid (Figures 1.28 and 3.2).

Phylum ANNELIDA (about 8 800 species)

The body is segmented, round or oval in section and is composed of three layers of cells: ectoderm, endoderm lining the cavity of the gut and mesoderm in between, forming muscles, reproductive organs, excretory organs, and the walls of blood vessels. The gut has a mouth and an anus and gut musculature moves the contents towards the posterior end. Blood, often containing red pigments, circulates in a closed blood system. There is a coelom between the muscles of the outer wall of the gut and those of the body wall, usually broken up into compartments by partitions called septa and opening to the exterior by tubes called coelomoducts. Externally, the segments are usually marked by grooves that encircle the body and, except at the anterior end and the extreme posterior end, the segments are identical, with serial repetition down the body of structures such as muscles, nerve ganglia, peripheral nerves, blood vessels. The body is covered with a cuticle; bristles called chaetae are often present, arranged segmentally.

Class Polychaeta

Typically marine worms, with chaetae carried on parapodia (lateral flaps of the body, one pair per segment). Usually there is a head bearing appendages. The reproductive

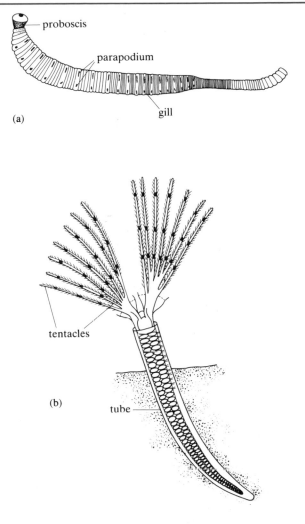

Figure A.28 Polychaeta. (a) The lugworm, *Arenicola*, drawn with dorsal surface down. (b) The fanworm, *Sabella* in its tube; some tentacles have been omitted.

organs are usually simple and the sexes are separate. Habits range from active crawling or swimming to burrowing or living in tubes. *Nereis* (ragworm; Figures 1.21 and 1.22 and Plate 10a) is an active worm that crawls by movements of the parapodia and swims by lateral undulations of the body. Food is seized by the everted pharynx. *Arenicola* (lugworm; Figure A.28a) burrows in muddy sand. *Sabella* (fanworm; Figure A.28b) lives in a tube and feeds on small particles collected by cilia on the tentacles that are extended as a 'fan' when the worm pushes its head out of the tube.

Class Oligochaeta

Typically terrestrial or freshwater hermaphrodite annelids, that burrow in the earth or swim or burrow in freshwaters. Without a distinct head or parapodia but with a few chaetae and complex reproductive organs. *Lumbricus* (earthworm; Figure A.29).

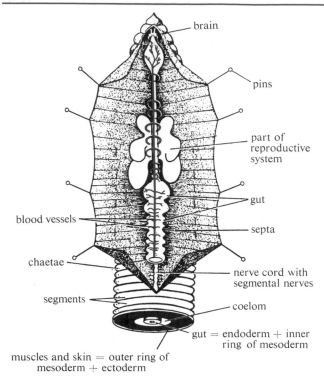

Figure A.29 Oligochaeta. A dissection of the anterior end of the earthworm, *Lumbricus*.

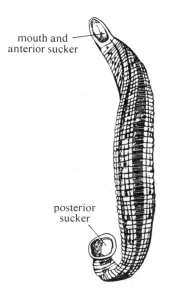

Figure A.30 Hirudinea. The leech, *Hirudo*, now rare but formerly common in ponds. It sucks the blood of mammals, including humans.

Class Hirudinea

Body usually flattened and with 32 segments, but several external grooves per segment. Two suckers, one surrounding the mouth and the other at the posterior end, without parapodia and usually without chaetae. Hermaphrodite, with complex reproductive organs. Coelom restricted by development of a cellular mass called botryoidal tissue. Leeches occur in marine, freshwater, and terrestrial habitats. Many are external parasites, sucking blood from animals to which they attach themselves with the suckers; others are carnivores or detritus feeders. *Hirudo* (medicinal leech, Figure A.30).

Phylum POGONOPHORA (about 45 species)

Long, very slender worms (Figure 2.1), mostly tube-dwellers in deep oceans; with one or more tentacles at the anterior end. Ectoderm and mesoderm are present but endoderm and gut are absent. There is a single anterior coelomic cavity and two pairs of posterior coelomic cavities and a closed blood system; the ventral vessel is enlarged to form a heart surrounded by a pericardial sac; blood flows forwards ventrally and backwards dorsally and the nervous system is simple.

Phylum ARTHROPODA (well over 1 000 000 species)

The bilaterally symmetrical, segmented body is enclosed in a chitinous exoskeleton, usually rendered impermeable (e.g. by oily secretions in insects) and hard by additives (such as calcareous deposits in crabs). The life cycle involves many stages, often including larval forms, separated by moulting of old exoskeleton. The body is usually divided into regions (e.g. head, thorax, and abdomen of insects) and there is normally one pair of jointed appendages per body segment, that function as mouthparts, respiratory structures, locomotory limbs or sense organs. The gut has a mouth and an anus. The main body cavity is a haemocoel but there are coelomic cavities in embryos. With a few exceptions, cilia are absent.

There is a great diversity of form within the phylum and much controversy about the major subdivisions and about whether certain small groups should be included, e.g. tardigrades, linguatulids. Several formerly abundant groups are completely extinct, e.g. trilobites (Figure 2.11). Living species are classified into three subphyla, Crustacea, Chelicerata and Uniramia.

Subphylum CRUSTACEA (about 26 000 species)

Body often divides into a head plus thorax and an abdomen (Figure 1.24); exoskeleton is stiffened with calcium salts. Appendages typically biramous (two major branches), usually one or two anterior pairs of sensory antennae, in front of the third pair of appendages which are mandibles used only in feeding that bite by fore-and-aft movements. There are usually one or more other pairs of feeding appendages, and at least four pairs of biramous appendages are locomotory appendages and may also carry gills. Most aquatic crustaceans have a planktonic, free-swimming larva called a nauplius (Figures A.33b and A.34b). Crustaceans are mostly marine and freshwater with a few terrestrial forms; a great diversity of habits from swimming to walking or burrowing and from feeding on minute particles to predation.

Class Branchiopoda

Free-swimming, usually of small size and often very abundant in the plankton of lakes; some are marine. Most common forms belong to the order Cladocera: usually with bivalve carapace over the body but head free; four to six thoracic segments with flattened appendages; large, biramous antennae, used in swimming; abdomen reduced, without appendages. The carapace forms a brood pouch that contains eggs. Many species consist only of parthenogenetic females, except at certain seasons when males appear and sexual reproduction occurs. *Daphnia* (water flea; Figure A.31a).

Class Ostracoda

Mostly small crustaceans common in the sea and freshwaters. Carapace includes body and head, so they look like small bivalve molluscs but swim rapidly and smoothly.

Class Copepoda

Typically small, free-swimming species in the sea and freshwaters but many are parasites of polychaetes, echinoderms, other crustaceans and fishes, particularly on the gills. The sexes are separate and there is a nauplius and several other (copepodid) larval stages. Planktonic copepods typically lack a carapace and swim jerkily with the antennae and six pairs of thoracic limbs, e.g. *Cyclops*. *Calanus finmarchicus* (Figure A.32) is one of the most abundant animals in the plankton of temperate and arctic oceans. The long antennae and two other pairs of head appendages propel the animal slowly forwards and produce feeding currents. As the water passes through the long setae on the mouthparts, tiny suspended particles such as diatoms (Figure 1.6c, d and e) are trapped, then scraped into the mouth by bristles on other appendages. When not feeding, *Calanus* moves quickly and jerkily by vigorous movements of the paired appendages (Figure A.32).

Parasitic copepods include almost unmodified ectoparasites (with reduced abdomen and appendages used for clinging to the host) and highly specialized sac-like forms recognizable as copepods only by the production of typical egg sacs and eggs which hatch into nauplius larvae. The adult female *Chondracanthus* is just a sac attached to gills of fishes and the reduced male clings onto her body with hook-like antennae. *Xenocoeloma* as an adult is a mere mass of tissue lying in a sac formed by the epithelium of a polychaete worm, without any sign of appendages; it is hermaphrodite and produces two long egg-sacs; the eggs hatch into nauplius

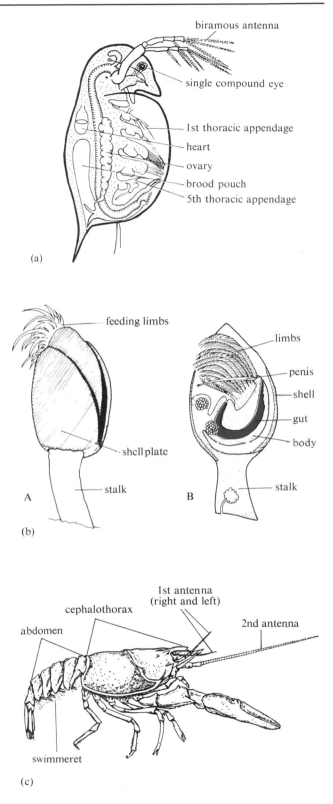

(a)

(b)

(c)

Figure A.31 Crustacea. (a) The cladoceran, *Daphnia* (magnified × 25). (b) The goose barnacle *Lepas*: (A) whole animal; (B) cut in half lengthways. (c) The freshwater crayfish, *Austropotamobius*.

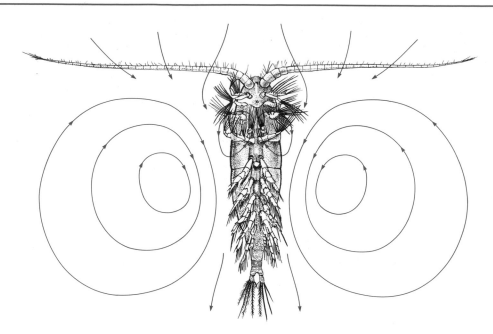

Figure A.32 *Calanus finmarchicus*, a marine planktonic copepod, swimming slowly, seen from below. The arrows show water currents created mainly by movements of the long antennae.

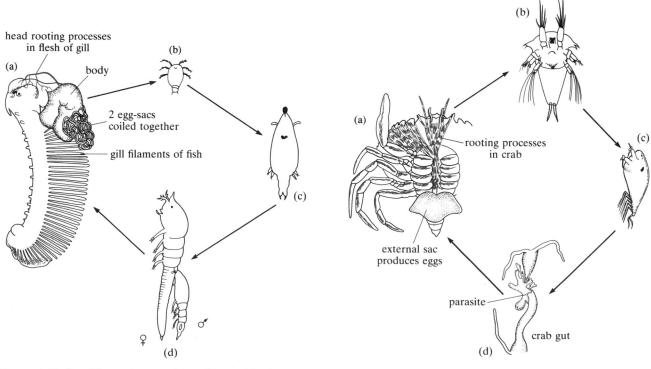

Figure A.33 Parasitic crustaceans (class Copepoda). *Lernaea*: stages in the life cycle: (a) adult female attached to gill arch of codfish; (b) nauplius larva hatches from egg and swims to a flatfish; (c) stage attached to flatfish gill; (d) male and female become free-living and mate; the males die at this stage and females attach themselves to gills of the cod family and grow to give the stage in (a).

Figure A.34 Parasitic crustaceans (class Cirripedia). *Sacculina*: stages in the life cycle: (a) a crab, *Carcinus*, infected with rooting processes and carrying sac with gonads; (b) nauplius larva hatches from egg; (c) cypris larva attaches itself to crab; (d) adult parasite grows inside the crab.

larvae. *Lernaea* (Figure A.33) has a complicated life cycle: there is a free-living nauplius larva that attaches to the gills of a flatfish. The adults become free-living again, but after mating the male dies and the females attach to the gills of fast-swimming fish such as cod where the eggs mature and are released.

Class Branchiura

Biramous thoracic swimming appendages; flattened body; pair of suckers; no egg-sac. All are ectoparasitic on marine and freshwater fishes. *Argulus* (fish louse).

Class Cirripedia

Adults are sessile or parasitic and usually hermaphrodite; larval stages free-swimming usually including a nauplius larva, with fronto-lateral horns bearing glands, which develops into a cypris larva which settles and metamorphoses into the adult. Barnacles (Plate 11h) have carapace with calcareous plates forming a shell, often closed by hinged plates; six pairs of biramous thoracic appendages form a cast-net. *Balanus*, *Lepas* (Figure A.31b).

Parasitic cirripedes may lack shell, appendages and gut and grow 'roots' penetrating tissues of host. *Anelasma* lacks filtering limbs and the body is partly embedded in the skin of sharks and has root-like processes by which it feeds on the shark's muscles. Adult *Sacculina* (Figures A.34a and d) are parasitic on shore crabs and consist of a simple sac under the crab's abdomen from which 'roots' penetrate through the tissue of the crab. It produces a typical nauplius larva that moults to form a cypris which settles on the crab and grows into it, rather like fungal hyphae.

Class Malacostraca

Body of 20 segments (six head, eight thoracic, six abdominal); varied appendages. Mostly large crustaceans found in seas, freshwaters, on land, and in hot springs and underground waters, e.g. crayfish (Figure A.31c), shrimps, lobsers, crabs, wood-lice. The two largest superorders are Peracarida and Eucarida.

Superorder Peracarida   Carapace reduced or absent; females with brood pouch; larval stages are generally absent. Mysids (kangaroo shrimps) have stalked eyes, biramous thoracic appendages and long abdomen. The most familiar orders are:

Order Amphipoda   Laterally compressed with sessile eyes, uniramous (one major branch) thoracic appendages; the majority live in shallow seas, some live in freshwaters. Sandhoppers and freshwater shrimps.

Order Isopoda   Dorso–ventrally flattened with sessile eyes, uniramous thoracic appendages; most live in seas, freshwaters and on land in damp places, e.g. slaters and woodlice (Figure 2.15). A few are parasites on other crustaceans.

The first larval stages of parasitic species resemble small free-living isopods but with piercing mouthparts; they swim and attach themselves to copepods. After several moults, the parasite leaves the copepod and seeks the second (final) host, often a prawn or crab. *Bopyrus* (Figure A.35) lives in the gill chamber of a prawn, sucking its blood; the first individual to settle becomes a female, which develops a large brood pouch (giving the prawn 'face-ache'), and any later ones become males. Other isopods are internal parasites and the females are maggot-like; males much smaller. *Danalia* is a parasite of *Sacculina* (itself parasitic on a crab); others are parasitic on fish either as adults (Cymothoidae) or as larvae (Gnathiidae).

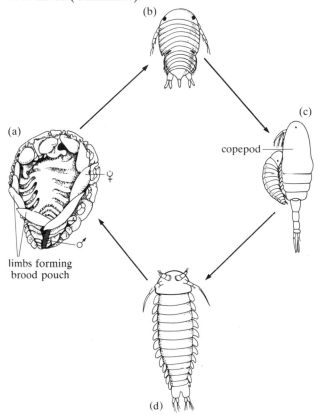

Figure A.35 Parasitic crustaceans (class Isopoda). *Bopyrus*: stages in the life cycle: (a) adult female carrying a dwarf male, taken from gill cavity of prawn; (b) epicaridium larva emerges from brood pouch and swims to copepod; (c) microniscus larva attached to copepod; (d) cryptoniscus larva leaves copepod and seeks a prawn.

Superorder Eucarida   Carapace covers thorax; eyes always stalked; no brood pouch and life cycle usually includes several swimming larval stages (Figures 2.5a and b): e.g. nauplius, megalopa.

Order Euphausiacea   With similar, biramous thoracic appendages and a long abdomen; most are marine filter-feeders, eating zooplankton, including *Calanus*. The Antarctic species *Euphausia superba* (krill) occurs in huge swarms and is the major food for many sea birds and large marine mammals, including the blue whale.

Order Decapoda Body form ranges from lobster-like (with long abdomen), to crab-like (with short abdomen tucked under thorax); hermit crabs have an asymmetrical abdomen and the last thoracic appendages are small compared with others. Mainly marine but include freshwater and a few terrestrial forms (but they must return to the sea to spawn). Shrimps, lobsters (Figure 3.8), crayfish (Figures 1.24 and A.31c), crabs (Plate 11g), hermit crabs.

Subphylum CHELICERATA (about 60 000 species)

Mostly small, terrestrial carnivorous or detritus-feeding arthropods (e.g. scorpions, spiders, mites) but *Limulus* (horseshoe crab) is marine and there are a few other aquatic and parasitic species. The body is divided (Figure A.36) into prosoma (with mouth and walking legs) and opisthosoma (usually with respiratory gill or 'lung books') and the exoskeleton is strengthened with a tanned protein. The appendages are primitively biramous (two major branches) but most are uniramous (one major branch) in terrestrial species; the most anterior pairs are chelicerae (usually used for grasping food) and pedipalps (usually sensory), never antennae (Figure 1.23d, e and f). The next four pairs of appendages are walking legs, some of which surround the mouth and bear jaw-like structures (gnathobases) on the basal segments, which chew by an in-and-out movement.

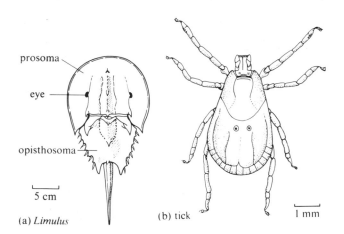

Figure A.36 Chelicerata. (a) *Limulus* (class Merostomata). (b) A tick (class Arachnida).

Class Merostomata (5 species)

Aquatic arthropods, with five pairs of powerful walking legs, paired chelicerae (claw-like feeding appendages) and several rows of ventral gills, a dorsal carapace and a long pointed tail (telson) at the posterior end. Palaeozoic members of this class, called eurypterids, were abundant and diverse, and some were very large, up to 3 m long. The living species are scattered but locally abundant. They live in coastal seas and feed on annelids that burrow in muddy sand. *Limulus* (horseshoe crab; Figure A.36a).

Class Arachnida (over 50 000 species)

Typically four pairs of walking legs and sensory or prehensile chelicerae and pedipalps. Almost all are terrestrial throughout the life cycle and breathe air through 'lung books' or tracheae. Mostly carnivorous; the principal predators on other arthropods. The four major living orders are:

Order Scorpionida Scorpions (Figure 1.23e) have an opisthosoma divided into a broader part and a narrow posterior part ending in a telson bearing a poisonous gland and forming the sting. The chelicerae are short and powerful and the pedipalps form large claws. There are four pairs of lung books. Scorpions prey mainly on other arthropods, especially large insects, but will attack vertebrates if disturbed; a few species are sufficiently poisonous to kill humans.

Order Araneae Spiders (Figure 1.23d, Plate 11c) have a globose opisthosoma attached to the prosoma by a narrow waist. The chelicerae are two-jointed and contain poison glands; the pedipalps resemble legs but are usually sensory (and used also for copulation by males). There are 'lung books', or tracheae, or both. Silk is produced in complicated 'spinnerets' derived from opisthosomal appendages; spider silk is remarkably strong but also elastic and forms the familiar webs, cocoons for eggs and other structures. All spiders are carnivorous, sucking up the partly digested juices of their prey; some chase prey, others burrow, make traps, or spin webs.

Order Opiliones Harvestmen (Plate 11d) resemble spiders but have no waist and have very long, slender legs. They feed mainly on arthropods but they may also eat carrion and bits of vegetation.

Order Acarina Ticks and mites (Figures 1.23f and A.36b; Plate 11b) have a broad junction between prosoma and opisthosoma. The chelicerae may be piercing or clawed and the pedipalps may be leg-like or clawed; tracheae and spiracles are present only in larger species. Most acarines are very small but they are abundant and diverse in soil, litter and mosses, in dried grain and in freshwaters, as herbivores on plants (red spider mite) and in plant galls. In Antarctica, mites are the largest fully terrestrial carnivores. Many others are ectoparasites of a wide range of animals including most terrestrial reptiles, birds and mammals (ticks and feather mites) and may transmit pathogenic viruses and bacteria.

Subphylum UNIRAMIA (probably more than 1 000 000 species)

Typically terrestrial but there are many insects in freshwater (usually as larvae) and a few marine species. The exoskeleton is strengthened with a tanned protein. The appendages are uniramous; the most anterior pair are sensory antennae, the mandibles beside the mouth bite with the tip of the appendages (not the base) and have an in-and-out movement; the walking legs are always behind the head region. The rest of the body may be divided into thorax and abdomen (in insects) or consist of many similar segments.

Class Diplopoda

Millipedes (Figure 1.23c; Plate 11f) have an almost cylindrical body with from 25 to more than 100 segments, most of which are double (two pairs of legs per segment). Head with one pair of antennae and two pairs of mouthparts; rest of body with legs, close together under the body. They breathe air through unbranched tracheal tubes opening near the legs. Most are able to roll into a ball or spiral. Most are herbivorous, living in dark, moist places and pushing into vegetation, logs or soil.

Class Chilopoda

Centipedes (Figure 1.23b; Plate 11e) have a flattened body, usually with 15 to 25 segments but some have up to 175 segments. Head with one pair of antennae and three pairs of mouthparts, next segment bears poison claws, following segments bear walking legs, attached to the sides of the body so not close together. They breathe air through branched tracheal tubes. They are typically active, swiftly moving carnivores living under stones and logs or in litter.

Class Insecta

Subclass Apterygota   (wingless insects)

Order Collembola (about 1 500 species) Small or minute animals in which the body is divided into head, thorax and abdomen but the mouthparts and legs differ from those of pterygotes. They have a springing organ on segment four of the six-segmented abdomen, hence the common name springtails. Collembolans are very widely distributed on all continents including Antarctica and are often numerous in soil and damp habitats.

Order Thysanura Bristle-tails are similar in general structure to pterygotes but without any sign of wings and have biting mouthparts and long, tapering bodies, a pair of long antennae and three long 'tails'. They continue to moult after maturity throughout a long life. Mainly live in litter, under stones and bark, and in houses, e.g. *Lepisma* (silverfish), *Thermobia domestica* (firebrat). One of the largest species, *Petrobius maritimus*, is common on rocky beaches.

Subclass Pterygota (winged insects)

Insects are the only invertebrates with wings (formed by outgrowths of the middle and posterior thoracic segments and present only in adults). In adults (Figure A.37a), the head bears one pair of antennae and three pairs of mouthparts; the thorax bears three pairs of legs and (usually) two pairs of wings; the abdomen is without legs but may bear reproductive appendages and tail-like structures (cerci) at the posterior end. All terrestrial and some aquatic insects breathe air through branched tracheae that ramify through the body (Figures A.37b and c); some aquatic larvae have gills and tracheae are reduced in very small species and parasitic forms. Blood normally lacks pigments and the heart is elaborate only in large insects. The nervous system (Figure A.37d) consists of a chain of ganglia linked by pairs of connectives. Most insect eggs have a tough, rigid, waterproof covering. Insects grow by a series of ecdyses (moults). The form assumed by an insect between each ecdysis is called an instar; the changes in form between instars is called metamorphosis. In most insects, the final instar is called the imago (adult) and is sexually mature. Immature stages may resemble adults except for the presence of fully developed wings (hemimetabolous or exopterygote insects) or may be very different from the adults with no external sign of wings (holometabolous or endopterygote insects) (Figure 2.8). The juvenile stages of exopterygote insects are often called nymphs; those of endopterygotes are called larvae and there is an extra stage, the pupa, within which there is extensive reorganization of the body and from which the imago emerges. Insects are very diverse in form and habit, and include the largest number of species of any class of organisms (more than one-third of all living species are insects).

Order Ephemeroptera (at least 1 500 species) Soft-bodied exopterygote insects that are short-lived as adults. Wide forewings, reduced hindwings, both held up vertically when at rest and present in the adult (imago) and previous instar (subimago); adults lack mouthparts and do not feed. The larva (nymph) has biting mouthparts, lives in water, usually has three 'tails' and paired abdominal gills. Mayflies.

Order Odonata (at least 5 000 species) Strongly flying, predatory insects with biting mouthparts and large eyes; two equal pairs of membranous wings. Exopterygote insects with aquatic nymphs with raptorial 'mask' and rectal or caudal gills. Dragonflies, damselflies.

The orders Ephemeroptera and Odonata are sometimes classified together in a separate group, Palaeoptera, on account of the simple wing hinges; all other winged insects are then put in another group, Neoptera.

Order Orthoptera (at least 11 000 species) Medium or large exopterygote insects usually with long hindlegs used in jumping; biting mouthparts. Typically forewing is sclerotized (darkened and stiffened) and hindwing is fan-like and membranous. Wings reduced or absent in some species and many have stridulatory and auditory organs. Widely distributed. Grasshoppers (Plate 11j), locusts, crickets.

Order Dictyoptera (at least 5 500 species) Medium or large exopterygote insects with biting mouthparts. Forelegs raptorial in mantids; otherwise legs similar to each other, used for running. Forewings sclerotized and hindwings membranous. Mainly in tropics and sub-tropics. Cockroaches, praying mantis.

Associated with Orthoptera and Dictyoptera are several small orders including: Dermaptera (earwigs), Isoptera (termites or white ants), Phasmida (stick insects), and Plecoptera (stone-flies, with aquatic nymphs, often with thoracic gills and two 'tails').

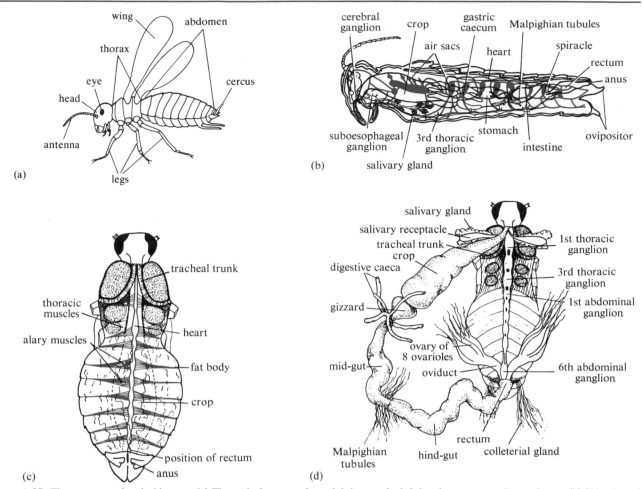

Figure A.37 The structure of typical insects. (a) The main features of an adult insect; the left-hand structures only are shown. (b) Side view of a locust to show the relative position of the gut and nervous system. The airsacs and main branches of the tracheal system are shown in red, superimposed on the other organs. (c) A cockroach with the dorsal exoskeleton removed to show the heart and fat bodies; thoracic muscles, cut across, are stippled. (d) A female cockroach, seen from above, dissected to show the gut, reproductive system and nerve cord.

Order Hemiptera (at least 50 000 species) Exopterygotes with piercing and sucking mouthparts of characteristic form. Wings variable but forewings often wholly or partly sclerotized. Wide distribution; many are pests or transmit diseases, e.g. aphids, scale-insects, plant bugs, bedbugs, water boatmen, cicadas. Scale-insects have instars that resemble the pupal instar of endopterygote insects. Sometimes divided into two orders.

Associated with Hemiptera there are several small orders including: Anoplura (sucking lice; Figure 2.23); Mallophaga (bird lice); Psocoptera (book lice); Thysanoptera (thrips).

Order Lepidoptera (at least 105 000 species) Endopterygote insects usually with a dense covering of scales on body, legs and wings; two pairs of membranous wings. Mouthparts normally lack mandibles; a suctorial proboscis often present. Larvae are caterpillars, usually with pro-legs on the abdomen. Butterflies and moths (Plates 7c, 11k, 11l).

Order Trichoptera (at least 3 600 species) Aquatic larvae that often live in cases; adults with hairy wings and antennae. Caddis flies.

Order Diptera (at least 95 000 species) Endopterygote insects in which hindwings form clubs (halteres) of sensory function, forewings are membranous. Maggot-like larvae without true legs; pupae either free or in a puparium (larval cuticle). Mouthparts usually piercing or sucking; most adults are diurnal and feed on nectar or decaying organic matter but many prey on other arthropods or suck vertebrate blood and may transmit disease. Mosquitoes, flies (e.g. *Drosophila*, Figure 1.25), daddy longlegs (Plate 11i).

Two small orders, Mecoptera (scorpion flies) and Siphonaptera (fleas), have characters in common with Diptera, Lepidoptera and Trichoptera.

Order Coleoptera (at least 400 000 species) Minute to large endopterygote insects (Figure 1.23a, Plate 11a, m, n) with biting mouthparts and the forewings modified as horny 'elytra' that meet edge to edge at rest and cover the folded hindwings and abdomen. Beetles are the largest order in the animal kingdom and very varied in habits (some are pests). Scarabs (Plate 11a), weevils (Plate 11m), ladybirds (Plate 11n).

Several small orders including Neuroptera (lacewings, ant-lions, and alder-flies) have characters in common with Coleoptera.

Order Hymenoptera (at least 200 000 species) Endopterygote insects with two pairs of membranous wings, hindwings smaller than forewings and interlocked with them by hooklets. Mouthparts primitively biting, often adapted for lapping or sucking. Abdomen with a 'waist'; females with ovipositors that may pierce, sting, or saw. Sawfly larvae resemble caterpillars; other larvae are legless maggots. A large order of specialized insects, many beneficial to humans and a few harmful. Many species are parasites on other insects. Some species have complex social organizations. Sawflies, ichneumon flies, bees (e.g. honey-bee, Figure 2.24), wasps, ants.

Phylum ONYCHOPHORA (about 70 species)

Soft-bodied animals with chitinous exoskeleton and hollow simple appendages operated by muscles working against the haemocoelic hydroskeleton, one pair of antennae, one pair of small, hard jaws, many pairs of walking legs. Body with many spiracles, each opening into a tuft of short unbranched tracheae; hence the animals are restricted to damp conditions such as decaying logs. *Peripatus* (Figure 2.14 and Plate 12).

Phylum ECHINODERMATA (about 6 000 species)

The body is unsegmented and there is no head. Typically the mouth (oral surface) is ventral and the anus (aboral surface) dorsal and the body has pentamerous (five-rayed) symmetry about the oral–aboral axis. Tube-feet and a water-vascular system (derived from the coelom) are found only in echinoderms and take part in locomotion, feeding, respiration and various other functions. There is no blood system and the nervous system consists of nerve rings, radial nerves, nerve nets and nerves to tube-feet and spines. Simple sense organs only. There is typically an internal skeleton of calcareous ossicles that may fit together to form a test (as in sea-urchins), or remain as separate small ossicles.

All living echinoderms are marine and produce large numbers of small eggs that hatch into minute ciliated larvae (Figure 2.7) that live in the plankton. The larvae are bilaterally symmetrical but change as they grow larger into pentamerous forms.

There are many fossil echinoderms. The living species are classified into five classes:

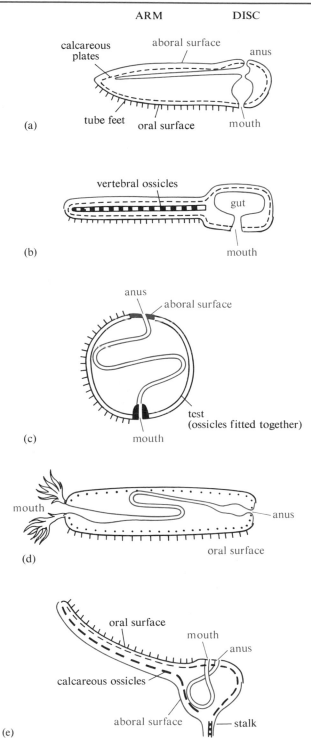

Figure A.38 Echinodermata. (a) The basic anatomy of a starfish, cut vertically through one arm (on the left). (b) The basic anatomy of a brittle-star, cut vertically through one arm (on the left). (c) The basic anatomy of a sea-urchin, cut vertically. (d) The basic anatomy of a sea-cucumber, cut vertically to show tube-feet below. (e) The basic anatomy of a sea lily (crinoid), cut vertically to show an arm on the left with the stalk cut short.

Class Asteroidea

Typically five-rayed with the arms not sharply marked off from the disc. They move by tube-feet usually on the sea-bottom but may burrow in sand. They may feed on bivalve molluscs or, by cilia, on small particles. Starfish (Figures 1.29 and A.38a).

Class Ophiuroidea

Typically five-rayed with the arms sharply marked off from the disc. They feed on small organisms or particles and move by horizontal bending movements of the arms which have a series of central ossicles that are sometimes called 'vertebrae'. Brittle-stars (Figures 2.12 and A.38b).

Class Echinoidea

The oral surface is greatly extended and the aboral surface reduced to a small area round the anus. The skeleton forms a test without arms. They move by tube-feet and spines. Spherical sea-urchins have five teeth and a complicated jaw-like structure called 'Aristotle's lantern' with which they scrape algae off rock surfaces. Heart urchins are bilaterally symmetrical with the anus 'posterior' and the mouth slightly 'anterior'; they burrow and feed on deposits; they lack Aristotle's lantern. Sea-urchin (Plate 14a and Figure A.38c), heart urchin, sand-dollars (Figure 2.4).

Class Holothuroidea

Oral surface greatly extended and aboral surface reduced; they usually live with the elongated oral–aboral axis horizontal. Body wall muscular with skeleton represented by small separate plates only; no spines. Tube-feet numerous, forming ten branched 'tentacles' round the mouth; no ampullae in the water-vascular system. They feed on small organisms or particles in deposits or in the water above. Sea-cucumber (Figure A.38d and Plates 14b and c).

Class Crinoidea

Normally live with oral surface facing upwards attached to the substratum by aboral apex of body either permanently or during development. Arms sharply separate from disc and usually branched, bearing pinnules. Tube-feet, without ampullae, function in feeding on minute particles and in respiration. Feather stars swim by lashing arms; sea lilies (Figure A.38e) have long stalks.

Phylum HEMICHORDATA (about 100 species)

The body is unsegmented but is divided into three regions. The gut has a mouth and anus. There is a blood-vascular system and separate coelomic compartments in the three regions of the body. The nervous system is very variable; there may be mid-dorsal and mid-ventral nerves. Most

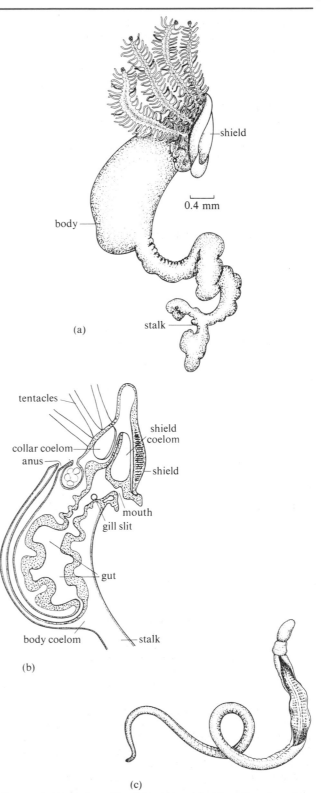

Figure A.39 Phylum Hemichordata. (a) Class Pterobranchia: *Cephalodiscus*, whole animal. (b) Longitudinal section of *Cephalodiscus* to show body cavities. (c) Class Enteropneusta, *Balanoglossus*, whole animal (up to 30 cm long).

species have gill slits connecting the pharynx with the sea outside. Some species produce a larva which is very similar to some echinoderm larvae.

There are two very different types of hemichordate:

Class Pterobranchia

Small or very small colonial animals living in the sea. They secrete tubes around themselves and feed on small suspended particles with the ciliated tentacles and U-shaped gut. *Cephalodiscus* (Figures A.39a and b).

Class Enteropneusta

Solitary worms of medium length with a proboscis (instead of shield and tentacles); many gill slits; straight gut; all marine. *Balanoglossus* (acorn worm) (Figure A.39c).

Phylum CHORDATA (about 54 000 species)

Coelomate animals with segmented muscles, a blood-vascular system and a mouth and an anus. Typically the body extends behind the anus as a tail. The nerve cord is hollow and dorsal to the notochord, which is dorsal to the gut; the pharynx is pierced by gill slits. There is an endoskeleton composed of cartilage and, usually, also of bone. Some of these features are present only in certain stages of the life cycle (e.g. humans have gills only briefly during fetal development).

The phylum is divided into three subphyla. The Cephalochordata and Urochordata are sometimes called Protochordates and contrasted with the Craniates (Vertebrates).

Subphylum CEPHALOCHORDATA (two genera and about 25 species)

The body is asymmetrical, with the segmented muscles and nerves alternating on the right and left sides. The notochord extends forward to the tip of the snout and consists of bands of muscle running dorso–ventrally, bound together by a sheath of cartilage. When the muscles contract (e.g. during swimming), they make the notochord stiff. The pharynx is very long, with up to 200 gill slits and is supported by endoskeletal bars of cartilage and the walls of the gill slits are ciliated. The cilia waft water in through the mouth, through the pharyngeal slits where it is filtered by a mucus sheet and out into the atriums, a cavity formed by folds of body wall fusing below the pharynx. The muscular notochord and long pharynx are unique features. The fins are not supported by hard skeletal tissue. There is a system of blood vessels, some with contractile walls, and the blood circulates forwards ventrally and backwards dorsally. Heart and blood capillaries are absent. There is a small planktonic larva. The adults (Figure 2.10) live partly buried in sandy gravel, feeding by filtering particles from the water above. They can swim by lateral undulations of the body. All are small (about 5 cm long) and marine. *Branchiostoma* (amphioxus) and *Asymmetron*.

Subphylum UROCHORDATA (sometimes called Tunicata; about 1 300 species)

The pharynx is very large with many gill slits which open into an atrium, as in Cephalochordata. Feeding is by filtering particles from the stream of seawater created by cilia on the pharynx. There is a heart, lying in a pericardium, and blood vessels; the heart reverses the direction of its beat at intervals. A notochord and segmented muscles are present only in the larva. The adult nervous system consists of a solid ganglion, close to the pharynx, and peripheral nerves. Tunicates are hermaphrodite and often multiply asexually by budding, forming colonies. Sexual reproduction includes a tadpole-like larval stage.

Class Ascidiacea

The adult (Figure 2.9a) consists of a large pharynx perforated with numerous gill clefts that function in filter feeding and respiration. Viscera and gonads are at the base. Adult sea-squirts are bottom-living marine animals, either solitary or forming colonies; the tadpole-like larva (Figure 2.9b) does not feed but may swim briefly in the plankton before settling on the bottom and metamorphosing into a sessile adult. *Ciona*, *Botryllus*, *Styela*.

Class Thaliacea

Adult salps resemble adult sea-squirts but are planktonic throughout their lives; many swim actively. The barrel-shaped body is surrounded by circular bands of muscle and perforated by numerous gill slits. *Salpa*, *Doliolum*.

Class Larvacea

The body form of the ascidian tadpole-like larva is retained throughout life. *Oikopleura* (Figure A.40a) is about 4 mm long and has a single large gill opening on each side and filter-feeds by means of a 'house' (Figures A.40b and c), i.e. a thin, transparent envelope secreted by the animal which it inflates by drawing water in by beating its tail. The house has two large openings covered with a mesh of threads which act as a coarse filter and prevent all but very small particles from entering. Within the house are two conical nets with very fine mesh which retain minute organisms (approx. 20 μm in size) which pass down the nets into the animal's mouth. The water current, created by the undulations of the animal's tail, passes through the filter-nets and out through a hole in the house. The nets are very fragile and are readily damaged or become clogged up with particles. A new house can be secreted in less than 30 minutes and may be replaced several times each day. *Oikopleura* houses filter smaller particles than any mesh so far manufactured by human efforts, including unicellular algae down to 1 μm in diameter. *Oikopleura* is sometimes very common in surface waters round the British Isles.

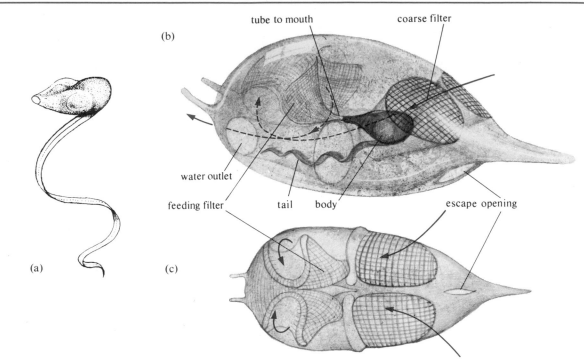

tube to mouth

coarse filter

(b)

water outlet

feeding filter   tail   body

escape opening

(a)

(c)

Figure A.40 Subphylum Urochordata, class Larvacea, *Oikopleura vanhoffeni*. (a) Animal removed from its house. (b) Animal in its house, side view. (c) Median view of house. The red arrows indicate the direction of water flow.

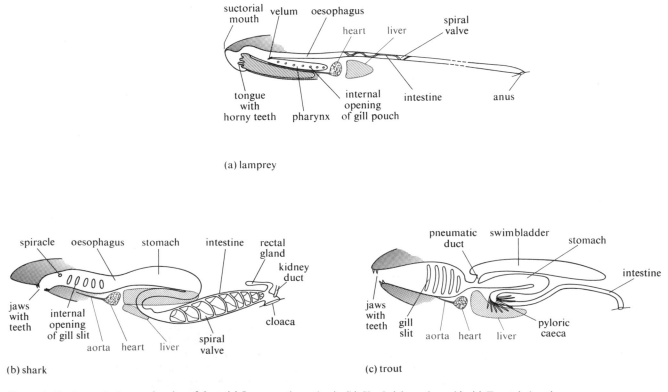

suctorial mouth   velum   oesophagus

heart   liver

spiral valve

tongue with horny teeth   pharynx   internal opening of gill pouch   intestine   anus

(a) lamprey

spiracle   oesophagus   stomach   intestine   rectal gland

kidney duct

jaws with teeth   internal opening of gill slit   aorta   heart   liver   spiral valve   cloaca

(b) shark

pneumatic duct   swimbladder   stomach

intestine

jaws with teeth   gill slit   aorta   heart   liver   pyloric caeca

(c) trout

Figure A.41 Gut and viscera of various fishes: (a) Lamprey (agnathan). (b) Shark (elasmobranch). (c) Trout (teleost).

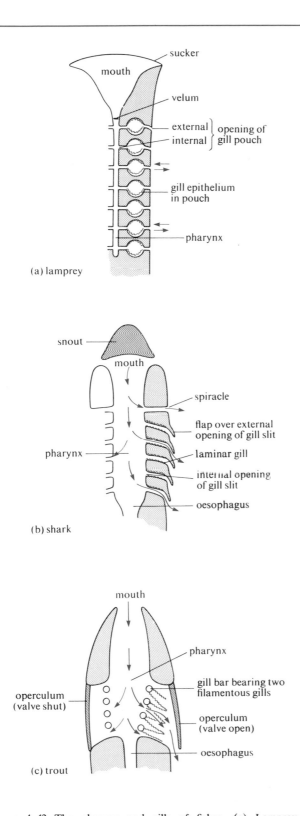

Figure A.42 The pharynx and gills of fishes. (a) Lamprey (agnathan). (b) Shark (elasmobranch). (c) Trout (teleost). Details are shown on right side only; gill epithelium shown in red and respiratory current shown by red arrows.

Subphylum CRANIATA (also called Vertebrata)

Vertebrates have the basic chordate characters (Figure 1.30). The anterior end of the body forms a head that carries the mouth, the brain (formed by expansion of the anterior end of the nerve cord) and special sense organs (vision, olfaction, balance and (usually) hearing). The brain and some sense organs are partly or wholly enclosed in part of the skeleton, the skull. Some general features of fishes are shown in Figures A.41–A.43. The notochord is usually replaced entirely or partly by bone or cartilage. Although cartilage is also found in some invertebrates, e.g. limpets (molluscs), bone is unique to vertebrates; vertebrate teeth contain dentine and enamel, very hard forms of bone that do not contain entire living cells and have more mineral and less protein than typical bone. The heart consists of three or more chambers (Figure A.44a), lies ventral to the gut, behind the head; blood flows forwards ventrally, then through the gill region and backwards dorsally (Figure A.44b). The kidneys are on the dorsal wall of the abdominal cavity (main coelom).

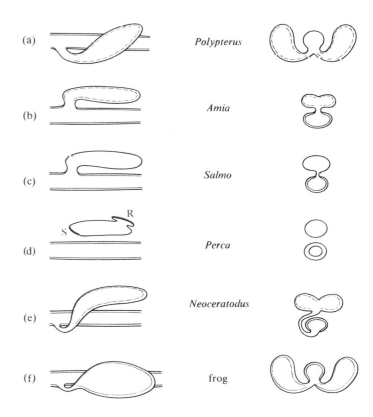

Figure A.43 Pharyngeal pouches in osteichthyan fishes and a typical amphibian, seen from the right side with the pharynx shown as a tube (on the left) and in section (on the right). S, area secreting oxygen; R, area resorbing oxygen. Red indicates epithelium where exchange of gases occurs between blood and air. (a) *Polypterus* is a chondrostean and (b) *Amia* (bowfin) is a holostean (Actinopterygii); (c) *Salmo* (salmon) and (d) *Perca* (perch) are teleosts (Actinopterygii); (e) *Neoceratodus* (lungfish) is a dipnoan (Sarcopterygii); (f) frogs are amphibians.

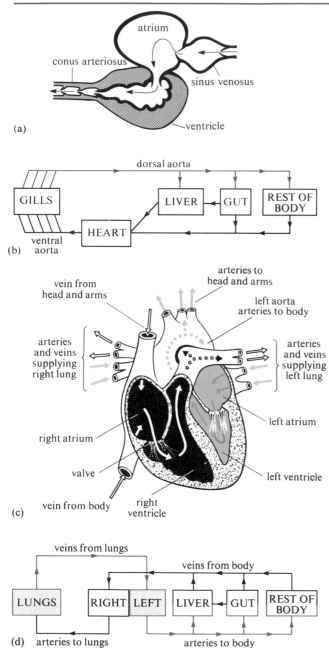

(a)

(b)

(c)

(d)

Figure A.44 (a) and (b) Fish hearts and circulation. (a) Elasmobranch heart. (b) Diagram of the single circulation of elasmobranchs and teleosts. Oxygenated blood is shown red; deoxygenated blood is black. (c) and (d) Mammalian heart and circulation. (c) A mammalian heart cut open lengthways and viewed from the front. The vessels leading to and from the heart are shown cut close to it; and arrows show the direction of blood flow. The cavities of the right side of the heart are coloured black with white arrows to show the direction of blood flow through them. The cavities of the left side of the heart are coloured grey with red arrows showing the direction of blood flow through them and the vessels connected with them. (d) Diagram of the double circulation of mammals. The boxes labelled 'right' and 'left' represent the two sides of the heart. Oxygenated blood shown red; deoxygenated blood shown black.

Class Agnatha (about 9 species)

The earliest vertebrates were agnathans (jawless fishes). There are a few living species (some locally abundant) that are very different in form and structure from the Palaeozoic agnathans. In the two modern groups (Figure A.45), the body is elongated (eel-like) and there are no paired fins or scales. The cartilaginous skeleton includes a simple skull and supports for the gill pouches, which open from the pharynx (*Myxine*) or from a blind duct below it (*Petromyzon*; Figure A.41a). The mouth is round (hence the common name 'cyclostome'). The notochord persists throughout life and peg-like cartilages may form beside it. The blood system, nervous system and kidneys are similar to those of jawed vertebrates, with minor modifications.

Petromyzon and other lampreys (Figure A.45a) are usually ectoparasites on fishes as adults and attach themselves by a suctorial mouth and rasp tissue with a horny tongue. They breed in freshwaters; the egg hatches into a filter-feeding ammocoete larva that lives buried in mud and metamorphoses, after several years, into an adult which may migrate to the sea. *Myxine* and other hagfishes (Figure A.45b) are marine, with reduced eyes and peculiar dental plates; they are mainly scavengers on dead fish and invertebrates. They do not have a larval stage.

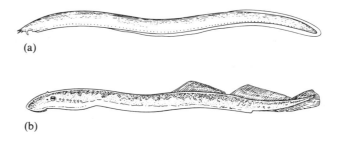

(a)

(b)

Figure A.45 Class Agnatha. (a) Lamprey, *Petromyzon*. (b) Hagfish, *Myxine*.

Class Chondrichthyes (about 570 species)

Bone is absent but the cartilaginous skeleton may be calcified (and consequently very hard). The structure of the scales is identical to that of the teeth of other vertebrates. There are usually two sets of paired fins (pectoral and pelvic). The intestine contains a spiral valve (Figure A.41b) but the other viscera are similar to those of other vertebrates. Fertilization is usually internal and males typically have claspers (intromittent organs); the eggs are large and yolky and some species are viviparous. Cartilaginous fishes are known as fossils from the Upper Devonian Period onwards (Figure 2.17). Most living and extinct species are marine.

Subclass Elasmobranchii (about 550 species)

Upper jaws that can move in relation to the skull. There are a pair of spiracles and five to seven pairs of gill slits, each with a separate external opening (Figure A.41b).

Superorder Selachii   The two large orders are Galeoidea and Squaloidea. Galeoidea are the typical sharks (Figures 1.33b and d and Plate 15a): generally active, round-bodied and good swimmers, either carnivores attracted by smell, or filter-feeders. The spiracles are small or absent; there are no spines in front of the dorsal fins and only one anal fin. Squaloidea (with spiracles, a spine in front of the first dorsal fin and no anal fin) are often flattened and many live near the bottom.

Superorder Batoidea   The body is typically flattened with a dorsal spiracle, gill slits ventral to the greatly enlarged pectoral fins and no anal fin. The teeth usually form pavements suitable for crushing molluscs. Skates (Figure 1.33c) and rays, guitar and saw-fishes, electric rays (*Torpedo*), sting-rays and eagle rays (*Manta*, Figure 1.33a).

Subclass Holocephali

These differ from elasmobranchs in the relationship between jaws and skull and in the details of the gill system, blood system and nervous system. Living species comprise a few genera of rare marine deep-water fishes, e.g. *Chimaera*.

Class Osteichthyes   (more than 30 000 species)

In addition to the basic vertebrate characters, there are pectoral and pelvic paired fins and the body is usually covered by bony scales. The pectoral fins are associated with a bony girdle which is connected to the back of the skull. Young fish have a cartilaginous skeleton, parts of which are later replaced by bone. A pair of pharyngeal pouches function as lungs or are modified (Figures A.43a–e). The gills are protected by a bony operculum (Figure A.42c). Fertilization is usually external and claspers are never present. There is usually a larval stage. The two major living subclasses are Sarcopterygii and Actinopterygii.

Subclass Sarcopterygii   (fleshy-finned fishes)

The paired fins have a central axis of bones and muscles, the tail has a webbed part dorsal to the vertebral column and the scales over the body have only a thin layer of enamel (or none). Sarcopterygii include two groups that are distinct from their first appearance in the fossil record (Figure 2.17). The Crossopterygii include many extinct forms and one modern species, *Latimeria chalumnae* (coelacanth; Figure 2.19) which lives off rocky submarine cliffs in deep water off the Comoro Islands (Indian Ocean). The Dipnoi include the lungfishes and the extinct ancestors of the tetrapod vertebrates that were the most numerous freshwater fishes of the Devonian Period. The three living genera of lungfish, *Neoceratodus* (Australia), *Protopterus* (Africa) and *Lepidosiren* (South America) are eel-like in shape (Figure 2.18) and have an unconstricted notochord and reduced ossification of the skeleton. They have lungs (Figure A.43e) and breathe air but also have gills; *Neoceratodus* requires access to oxygenated water, the others require access to air for survival. They have tooth plates and and powerful jaw muscles and crush molluscs and other animals. *Protopterus* and *Lepidosiren* males guard their eggs which hatch into tadpoles with external gills and suckers.

Subclass Actinopterygii   (at least 30 000 living species)

Bones and muscles do not extend beyond the base of the paired fins, the tail never has a webbed part dorsal to the vertebral column and in typical species the scales over the body have a thick layer of enamel. The great majority of living fish species are actinopterygians.

Superorder Chondrostei   Many extinct forms (Palaeonisciformes) (Figure 2.17) and two living genera. *Polypterus* lives in African tropical freshwaters. It breathes air through a pair of air sacs connected ventrally to the pharynx (Figure A.43a) and has unique pectoral fins. In the sturgeons, *Acipenser*, ossification and scales are much reduced and the mouth is highly modified. There is a dorsal airbladder with a wide duct connecting it to the oesophagus. Sturgeons breed in freshwaters but many spend much of their lives in the sea, where they grow to a length of several metres.

Superorder Holostei   Most of this group became extinct in the Cretaceous and Eocene Periods but there are two living genera in North American freshwaters. *Lepisosteus* (gar-pike), which may reach a length of 3 m and *Amia* (bowfin), which grows to about 70 cm, are carnivorous air-breathing fishes that have a bilobed airbladder with a duct connecting it dorsally to the gut (Figure A.43b).

Superorder Teleostei   Scales are generally thin and bony without enamel. The notochord is replaced by a vertebral column with bony vertebrae, expanded ventrally in the tail to form supports for the tail muscles and the externally symmetrical tail fin (Plates 15b and c). The maxilla (a bone in the upper jaw) is hinged anteriorly and free posteriorly and there are several different types of protrusible mouth. There are teeth on many bones of the mouth and pharynx. There is typically a single dorsal swimbladder (airbladder) which may be connected with the gut (Figures A.43c and d) and may secrete or absorb gas, allowing adjustments in buoyancy. Ducts from the ovary or testis lead directly to the outside without connection with the kidney or its ducts. Most teleosts produce numerous tiny eggs and a planktonic larva (Figures 2.5c and d). A few species are ectoparasites, mostly of other fishes.

The eight major groups of teleosts (Figure 1.31) are:

(i) Elopomorpha All have a leaf-like leptocephalus larva, e.g. tarpons and eels.

(ii) Clupeomorpha Mainly marine filter-feeders with special connections between skull and swimbladder, e.g. herrings and anchovies.

(iii) Osteoglossomorpha Tropical freshwater fishes of diverse forms, e.g. mormyrids.

(iv) Protacanthopterygii Mainly medium-sized carnivores, e.g. salmonids and many deep-sea fishes.

(v) Ostariophysi All have a set of small bones connecting the swimbladder with the ear and most have excellent hearing. The majority of freshwater fishes belong to this group, e.g. cyprinids (carps and minnows), catfishes, loaches, gymnotids (electric eel).

(vi) Atherinomorpha A small, diverse group, e.g. flying fishes, gars, guppies.

(vii) Paracanthopterygii Several fairly specialized orders that have similar jaw muscles, e.g. cod, angler-fishes, some deep-sea fishes.

(viii) Acanthopterygii With spiny fin-rays in at least the dorsal and anal fins (so may be painful to hold) and with a characteristic structure of the upper jaws. About half of all teleost species, e.g. perch, sunfishes, wrasse, blennies, gobies, tunas, flat-fishes, sticklebacks, trigger-fishes, puffer fishes (Figures 3.4 and 3.13).

Class Amphibia (about 2 000 species)

The adults typically have four legs (i.e. are tetrapods); the eggs and early stages often live in freshwaters. Amphibians have paired limbs (not fins) with digits. The pectoral girdle is not connected with the back of the skull. A middle ear is present with a single bone conducting vibrations from the tympanic membrane to the inner ear. The pelvic girdle is connected to the vertebral column by a sacral vertebra. Eggs are normally laid in water and the larvae are aquatic, typically having external gills. The adults (except some very small species and a few groups that do not metamorphose) breathe by a pair of lungs (Figure A.43f). Respiration takes place across the skin and lining of the mouth as well as through lungs. Both systemic arteries are functional in adults (Figure A.46a) and the ventricle of the heart is not divided. There are many fossil amphibians and three living orders.

Order Anura   Tailless forms which have enlarged hind-limbs and associated special features of the skeleton and musculature. Frogs and toads (Figure 1.34 and Plate 15e).

Order Urodela   Typically with long tails. Adults may be wholly terrestrial with no trace of gills but in some species metamorphosis from tadpole to adult is suppressed to different extents so that external gills are retained, limbs are reduced and the adults are aquatic. Newts and salamanders (Plate 15d).

Order Apoda   Body elongated, without limbs; tail very short or absent. Some have aquatic larvae, in others a 'larval' stage with external gills takes place inside eggs. Mostly tropical, burrowing species.

Class Reptilia (about 6 000 species)

Mostly terrestrial, quadrupedal vertebrates, without feathers, fur or mammary glands. The skin is keratinized (and almost waterproof) and may bear scales. Normally there are numerous teeth, replaced several times during life. Both systemic arteries are functional in adults (Figure A.46b) and oxygenated blood is incompletely separated from deoxygenated blood. Reproduction is by amniote eggs (Figure 2.20), which cannot survive in water (the embryo develops an amnion, chorion and allantois in addition to a yolk-sac). All chelonians, crocodilians and rhynchocephalians lay eggs in burrows on land; a few snakes and lizards are viviparous (the eggs are retained in the female's body until hatching). The young breathe air from hatching or birth and there is never a larval stage. Reptiles were abundant and diverse

during the Mesozoic Era and there are many fossil species, including the ancestors of birds and mammals. Most living species are tropical. The living orders differ greatly in adult anatomy.

Order Chelonia   The body is enclosed in a shell of bony plates firmly fused to the vertebral column and ribs. The skin over the shell forms horny plates (tortoiseshell). The pectoral and pelvic girdles lie within the ribs; the neck is very flexible. There are no teeth but the jaws bear horny plates and the soft parts are similar to those of lizards. Many are aquatic or semi-aquatic (including a few marine forms) but all lay eggs on land. Turtles (Plate 11h) and tortoises (Plate 15g).

Order Squamata   The skull of lizards and snakes is fenestrated and has other special features. The skin bears scales and sometimes dermal bony plates. Teeth are present on the jaws and other bones of the mouth. Most lizards are insectivorous or predators, but a few eat plants as adults. Most lizards (Plate 15h) have powerful legs but a few groups, e.g. *Anguis* (slow-worm) are legless. All snakes (Plate 15i and Figure 3.7) are limbless and all are carnivorous, having special features of the skull and jaws that enable them to swallow very large prey and also, in some families, allow them to erect the poison fangs. Snakes lack tympanic membranes and the eyelids are fused to form 'spectacles'.

Order Rhynchocephalia   A single living species on a few islands off New Zealand, *Sphenodon* (tuatara) is like a lizard in many ways but has unique features of the skull and skeleton. Rhynchocephalians were moderately abundant in the Mesozoic.

Order Crocodilia   Semi-aquatic, carnivorous reptiles that swim by tail movements and run on land. All lay eggs on land. The body is covered with bony plates with horny scales outside. Features of the skull, vertebral column and limbs reveal a close relationship to extinct dinosaurs and birds. Alligators and crocodiles (Plate 15f).

Class Aves (about 8 600 species)

Birds (Figure 1.35 and Plate 17) are homoiothermic vertebrates that lay shelled eggs on land (none is viviparous); most are able to fly. Feathers cover the body (scales occur only on the feet and/or face). The skull has a large braincase and large, efficient eyes; the jaws are without teeth but are enclosed in horny plates that form a beak. The heart consists of four chambers and the arterial and venous circulations are thus separate and the blood goes twice through the heart, once on its way to the lungs, and again before going to the rest of the body. The major blood vessel to the viscera and muscles (aorta) forms from the right systemic arch (Figure A.46d) and there is a complicated and very efficient respiratory system. Terrestrial locomotion is bipedal and the forelimbs form wings. There are many other skeletal features associated with flight including a large sternum (usually with a keel) from which flight muscles originate, a flexible neck and sacral vertebrae fused with pelvic girdle. There is a single (left) ovary and oviduct.

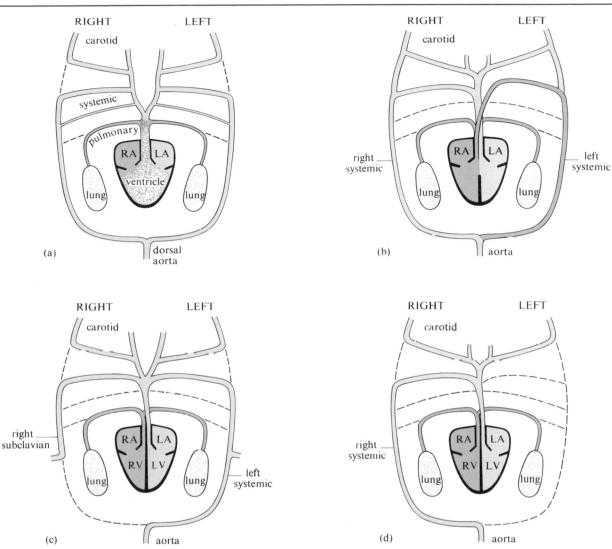

Figure A.46 Hearts and main arteries of tetrapod vertebrates. (a) Amphibian (salamander). (b) Reptile (lizard). (c) Mammal. (d) Bird. Red indicates oxygenated blood; black or grey is deoxygenated blood. Dashed lines indicate vessels that were probably functional in the fish ancestor and that are often present in embryos but absent in adults. RA, right atrium; LA, left atrium; RV, right ventricle; LV, left ventricle.

Birds are anatomically very homogeneous. There are two modern superorders:

Superorder Palaeognathae Flightless, mostly herbivorous birds called ratites (ostrich, rhea, emu, cassowary, kiwi) with a special type of palate (roof of mouth).

Superorder Neognathae All living birds except ratites with barbed feathers. Classified into at least 20 orders mainly on the structure of the beak (feeding adaptations), wings (types of flight) and legs (habits of life when not flying), including the 12 listed here (Figure 1.35).

Order Sphenisciformes Flightless, marine fish-eating birds, mostly on islands and coasts around Antarctica (Plate 17b). Penguins.

Order Procellariiformes Oceanic birds that come ashore only to breed, usually on remote islands. Powerful fliers and gliders, eating fish and planktonic invertebrates. Petrels, shearwaters and albatrosses.

Order Pelecaniformes Large, aquatic birds with all four toes webbed and a long beak with a pouch. Mostly feed by plunge diving. Pelicans.

Order Ciconiiformes Wading, mostly fish-eating birds with long, featherless legs and long necks. Herons, storks, ibises and spoonbills.

Order Anseriformes Swimming and diving birds in freshwaters and on coasts. Many species are mostly herbivorous. Ducks, geese and swans.

Order Falconiformes Birds of prey, with clawed feet, short, powerful, often hooked beaks and spectacular powers of flight. Eagles, buzzards, hawks, vultures.

Order Galliformes Mainly herbivorous birds of woodland or grassland that often run better than they can fly. Turkeys, fowl, pheasant, quails and ptarmigans.

Order Charadriiformes Omnivorous birds with long pointed beaks, living mostly on shores of seas and inland water. Waders, plovers, auks and gulls.

Order Columbiformes Herbivorous birds with four equal toes and powerful flight and that feed the young on a deciduous tissue of the throat ('pigeon milk'). Pigeons, sandgrouse and the extinct dodo.

Order Strigiformes Nocturnal, predatory birds with large, forward-facing eyes and feathers adapted for silent flight (Plate 17a). Most species hunt small mammals, particularly rodents. Owls.

Order Psittaciformes Seed- and fruit-eating birds with two toes pointing forwards and two pointing backwards and a stout, hooked beak (Plate 17c). Mostly social, brightly coloured forest-living birds, often highly intelligent and long-lived. Parrots, macaws, cockatoos and parakeets.

Order Passeriformes Song birds; all terrestrial, perching birds with four non-webbed toes, mostly eating insects and other small animals or seeds. The largest order: more than half of all known bird species, including sparrows, finches, swallows, robins, tits, crows.

Class Mammalia (about 5 000 species)

Fur normally covers the body but is sometimes restricted (scales may also be present, e.g. on a rat's tail) and there is normally an outer ear. The principal skeletal features are that the middle ear includes three auditory ossicles and the lower jaw consists of a single bone on each side. Teeth are found only on the dentary (lower jaw) and premaxilla and maxilla (upper jaw). Usually two sets are produced during life and may be of various shapes and relative sizes. A muscular diaphragm separates the thorax from the abdomen and assists respiratory movements. The heart (Figures A.44c and A.46c) consists of four chambers; no separate sinus venosus; two atria; two completely separate ventricles. The systemic and pulmonary (lung) circulations are thus separate (Figure A.44d) and the blood goes twice through the heart, once on its way to the lungs, and again before going to the rest of the body. The major blood vessel to the viscera and muscles (aorta) forms from the left systemic artery (Figure A.46c), the opposite arrangement to that of birds (Figure A.46d). The young are fed on milk secreted by female mammary glands for a period after birth (lactation). All eutherians and metatherians are viviparous. There are three living subclasses.

Subclass Prototheria (also called monotremes)

Many odd skeletal features, some of which are similar to those of reptiles. Females lay eggs in a nest or burrow and feed the hatchlings on large quantities of milk secreted from glands slightly different in structure from those of other mammals. Living species (Figure 1.36) occur in Australia, Tasmania and New Guinea. The duck-billed platypus is mainly aquatic and eats large invertebrates in lakes and rivers and nests in banks; the two species of echidnas (spiny anteaters) are terrestrial.

Subclass Metatheria (also called marsupials)

Marsupials (Figure 1.37) have various unique features of the skull and the skeleton. The young are born in a very immature state but with disproportionately large forelimbs; they crawl along the female's belly and attach to a nipple of a mammary gland (usually in a pouch). Lactation is prolonged and the composition of the milk changes as the young grows. No marsupials are fully aquatic. There are about 71 genera (258 species) grouped into several orders including:

Order Polyprotodontia Small, mostly forest-dwelling ominivorous and carnivorous marsupials with primitive teeth and feet. Most diverse in Australasia, but there are (and have been since the Cretaceous) some species in South America, a few of which have spread to North America. Opossums, numbat, bandicoots, Tasmanian devil, marsupial 'mouse'.

Order Diprotodontia Mostly herbivorous mammals with stout, flattened teeth; all native to Australia, New Guinea and surrounding islands. Wombats, koalas, phalangers (possums), kangaroos.

Subclass Eutheria (also called placentals)

The embryos are nourished in the uterus via placentae (Plate 16a) and are born at an advanced stage, sometimes able to run about (e.g. lambs, calves, elephants) but sometimes blind and helpless (e.g. kittens, rabbits). All neonates suckle milk (Plate 16b) for a period before progressing to the adult food. There are almost a thousand genera (including several partially and fully aquatic groups) classified into about 15 orders (Figure 1.38) including:

Order Insectivora (nearly 400 species) Small mammals with five toes on all limbs and numerous teeth. Mostly predators of insects and other invertebrates. Probably the closest of all living mammals to the ancestors of eutherians. Moles, shrews, hedgehogs.

Order Primates (over 200 species) Basically arboreal, mainly herbivorous mammals with large eyes and brains and elongated, flexible limbs. Lemurs, monkeys (Plate 16j and Figure 2.3), apes, humans.

Order Chiroptera (nearly 1 000 species) Insectivorous or frugivorous mammals in which the forelimbs (especially the elongated digits) support folds of skin that form wings used in active flight. Bats (Plate 16c), flying foxes.

Order Carnivora (nearly 250 species) With long stout canine teeth (Figure 2.2) and usually at least four clawed toes on each foot. Most are predators, e.g. dogs, wolves, polar bears, raccoons, ferrets, hyaenas, cats (Plate 16f), but some are partly or entirely herbivorous, e.g. brown bears (Plate 16b), badgers, giant panda.

Order Pinnipedia (34 species) Aquatic, predatory mammals which breed and raise their young on land or on ice. Similar in many ways to carnivores but with reduced limbs and ears and massive body musculature. Seals (Plate 16g), sea-lions, walruses.

Order Proboscidea (two species) Herbivorous mammals with a prehensile nose (trunk) and in which the very large molar teeth are worn sequentially rather than simultaneously. Living species have large ears and feet with four or five toes. Elephants (Plate 16k) are the largest living terrestrial mammals.

Order Cetacea (79 species) Wholly aquatic mammals that give birth and suckle under water. Some are toothless and filter zooplankton by baleen (modified hair), others retain teeth and feed on fish, squid or marine mammals. Whales, porpoises, dolphins.

Order Perissodactyla (17 species, some probably extinct in the wild) Non-ruminant, herbivorous mammals with one or three hooved digits on each foot; in modern horses, each foot retains a single hoofed digit. Tapirs, rhinoceros, horses.

Order Artiodactyla (nearly 200 species) Primarily herbivorous, running mammals usually with horns and two or four hooved digits in each foot. Suids (pigs and hippopotamuses) lack horns and have simple stomachs and a more omnivorous diet. The majority are grazing or browsing ruminants, with compound stomachs that harbour symbiotic micro-organisms. Camels, giraffes (Plate 16i), bison (Plate 16h), antelope and deer.

Order Rodentia (about 1 700 species) A single pair of upper and lower incisors form the characteristic gnawing teeth, which grow throughout life; the grinding molars also grow continuously. Most rodents are small, fast-breeding herbivores or omnivores, capable of eating hard material, particularly seeds. This order includes rats, mice, hamsters, guinea-pigs, squirrels, coypu.

Order Lagomorpha (65 species) Herbivorous mammals with many similarities to rodents but with two pairs of upper incisors and other skeletal differences. Pikas, hares, rabbits.

ANSWERS TO QUESTIONS

CHAPTER I

Question 1 **Proteins and DNA or RNA only.**

Question 2 Table 1.1 is completed as shown in Table 1.4.

Table 1.4 A comparison of prokaryotes and eukaryotes (completed).

	Prokaryotes	Eukaryotes
An outer membrane around the cell	+	+
Genetic material mainly confined to a membrane-bound nucleus	−	+
Organelles such as chloroplasts and mitochondria	−	+
Enzymes for aerobic respiration	+	+
Capacity for exchanging genetic material (sexual reproduction)	+	+
Capacity for cell division by mitosis	−	+

Table 1.5 Some features of the body plans of some animal phyla (Table 1.3 completed).

Phylum Class Common Name	Cnidaria		Platyhelminthes Turbellaria Planarian	Nematoda
	Medusa	Polyp		Roundworm
Symmetry				
radial (R), bilateral (B) or pentamerous (P)	R	R	B	B
Cavities in the body				
Gut or enteron	+	+	+	+
Pseudocoel	−	−	−	+
Coelom	−	−	−	−
Haemocoel or blood system	−	−	−	−
Segmentation of all or part of the body	−	−	−	−
Locomotory appendages (type)	−	−	−	−
Skeleton (type)	mesogloea		−	−
Anus present	−	−	−	+
Musculature				
Muscles in gut wall	+	+	−	−
Circular body muscles	+	+	+	−
Longitudinal body muscles	−	+	+	+
Muscles in locomotory appendage(s)	−	−	−	−
Any special features not already mentioned	cnidoblasts		eversible pharynx no anus	non-elastic cuticle

Question 3 (i) is shown by some bacteria and all cyanobacteria (b); (ii) is shown by some bacteria and some cyanobacteria (c); (iii) is shown by bacteria but never cyanobacteria (d); (iv) is shown by cyanobacteria only (e); (v) is shown by all prokaryotes (a); (vi) is shown by bacteria but never cyanobacteria (c); (vii) is shown by neither bacteria nor cyanobacteria (f).

Question 4 (a) Type D, e.g. *Ulva* and some land plants. There is alternation of haploid and diploid multicellular phases; spores are the products of meiosis and the gametes are produced by mitosis from the haploid gametophyte.

(b) Type C, in which the gametes arise directly from meiosis and there is no multicellular haploid phase.

(c) Only type D. None of the others has both multicellular diploid and multicellular haploid phases.

Question 5 (a) is a simple chlorophyte plant, probably an alga; (b) is a chain of cyanobacteria; (c) is a fungal colony.

Question 6 (a) Porifera: (i), (vi); (b) Cnidaria: (ii), (iv); (c) Platyhelminthes: (iii), (iv); (d) Nematoda: (iii), (v), (vii); (e) Annelida: (iii), (v), (vii), (viii), (ix); (f) Mollusca: (iii), (v), (vii), (x); (g) Arthropoda: (iii), (v), (x); (h) Echinodermata: (iii), (v), (vii), (xi), (xii); (i) Chordata: (iii), (v), (viii), (xi).

Question 7 Table 1.3 should be completed as shown in Table 1.5.

Annelida Polychaeta Ragworm	Arthropoda Crustacea Shrimp	Mollusca Gastropoda Snail	Mollusca Cephalopoda Squid	Echinodermata Starfish	Chordata Fish
B	B	B	B	P	B
+	+	+	+	+	+
–	–	–	–	–	–
+	+	+	+	+	+
+	+	+	+	–	+
+	+	–	–	⊤	+
parapodia	jointed limbs	–	fins	tube-feet	fins
–	exoskeleton	shell	reduced shell	calcareous plates	endoskeleton
+	+	+	+	+	+
+	+	+	+	+?	+
+	–	–	–	+?	+
+	+	+	+	+	+
+	+	–	mantle muscles	tube-feet	+
chaetae		head-foot visceral mass mantle		tail	notochord in embryo

Question 8 (a) Phylum Chordata, class Reptilia, probably a lizard or croco-dile-like animal.

(b) Phylum Arthropoda. Although most arthropods have pairs of jointed legs or mouthparts, some, such as maggots, are legless but have the other essential features of the phylum. Since it has a head, thorax and abdomen it is probably an insect.

(c) Tapeworm (Phylum Platyhelminthes, class Cestoda).

(d) Slug (Phylum Mollusca, class Gastropoda).

(e) Earthworm (Phylum Annelida, class Oligochaeta).

CHAPTER 2

Question 1 Examine the structure and identify its properties; observe the behaviour of the species in the wild and note the role of the structure in normal activities; compile information about the occurrence of similar structures in other species; observe the behaviour of specimens in which the structure is naturally absent or deficient, or in which the structure has been experimentally removed or modified.

Question 2 All the statements except (d) are teleological. They could be rephrased as:

(a) There is very little plant food in the Arctic in winter. Herbivorous bears are absent from the Arctic, but carnivorous polar bears prosper.

(b) The males are larger and have stronger teeth in many species in which the males challenge predators and rival males that approach females with which they are associated.

(c) In cats, retraction of the claws except during climbing or prey capture minimizes wear of the sharp points.

(e) Tall trees that have long, spreading roots are not blown down by strong winds.

(f) Animals such as newts that desiccate readily are active on land mainly at night.

(g) Sea-anemones have a large mouth and swallow large prey.

(h) Moss gametes are dispersed by rain.

Question 3 Only (a) and (c) are correct. Although larvae often feed on transiently abundant foods, e.g. caterpillars eat new leaves and shoots and maggots feed on carrion, there is no particular kind of food that is exploited exclusively by larvae (b). The presence of a larval stage has no direct bearing upon the evolutionary origins or abundance of a species (d and e).

Question 4 Only (c) and (d) are true. It is the basic body plan, not its adaptations nor the course of its development, that is the basis for identifying relationships, so (a) and (b) are irrelevant. The proportion of the life cycle that is spent as a larva is irrelevant to the interpretation of its structure, so (e) is false.

Question 5 (a) The gross anatomical structure is the most abundant and reliable kind of information. (b) Dead organisms may be displaced from their original habitat before fossilization and hence many assemblages do not reflect living communities of organisms. (c) The sources of food can only be established by interpreting the functional anatomy (see Section 2.1). Establishing likely predators depends upon identifying structures adapted to protection and finding prey and predator in the same habitat. (d) and (e) Timing of first appearance and extinction can only be estimated accurately for species with tough hard parts for which there is a high and constant probability of fossilization (e.g. estuarine or shallow marine species). Sometimes estimates of the time of extinction are wildly inaccurate. For example, although coelacanths have been absent from the fossil record since the Mesozoic, and they were therefore thought to be extinct, specimens were found alive for the first time in the Indian Ocean only 50 years ago.

Question 6 Lepidoptera, Gastropoda and Carnivora have undergone extensive adaptive radiation. There are numerous different kinds of butterflies and moths (and their caterpillar larvae) and gastropods have diversified in freshwater and on land as well as in the sea. Some members of the order Carnivora have become omnivores or herbivores, then secondarily carnivorous and the group includes species of a wide range of sizes (see Section 2.1.1). The other groups have not diversified, although so far as can be determined from comparative studies and the fossil record, all except humans have been in existence for a long time.

Question 7 All these features except wood occur in aquatic and terrestrial organisms. Skin, hard skeleton, lungs, limbs and jaws are present in one or more aquatic groups as well as in terrestrial related species. All these features occur in some fishes (e.g. lungfish) and hard skeletons, limbs and mouthparts also occur in aquatic arthropods (e.g. crustaceans). Although most primarily aquatic arthropods have gills, some secondarily aquatic freshwater insects have tracheae. Wood, flowers and leaves first appeared in terrestrial or semi-terrestrial plants, but only flowers and leaves are retained in species that have later become aquatic. Wood is absent from all algae and from bryophytes and fern-like plants. It does not form in fully aquatic seed plants such as waterlilies and rushes, although shore-living trees such as willows and mangroves are woody.

Question 8 Only (f) is generally true. Some mutualisms (e.g. lichens) are very ancient, so (a) is false. Symbionts often have different biochemical capacities from those of their free-living relatives, e.g. mechanisms for avoiding the host's immune response, but they are not necessarily simpler, so (b) is false. Many different phyla and kinds of micro-organisms are symbionts, so (c) is false. Both ectoparasites and endoparasites may have elaborate specializations to their way of life and be host-specific, so (d) is false. There is no evidence that parasitism promotes extinction, so (e) is false.

Question 9 (a) In plants: wood, impermeable covering on leaves and stems, seed formation; in animals: tracheae, mechanically stiff exoskeleton, exoskeleton impermeable to water.

(b) In plants: leaves containing toxic substances; in animals: biting mouthparts; sucking mouthparts.

(c) In plants: scented flowers, brightly coloured flowers; flowers that produce nectar; in animals; two pairs of wings, large eyes, sucking mouthparts.

CHAPTER 3

Question 1 (a) Acetylcholine was first shown to act as a neurotransmitter in the common frog. Frogs are poikilothermic so their tissues function well at room temperature; they have long hindlimbs and powerful leg muscles. Poisons from certain tropical tree frogs are used to investigate the properties of ion specific channels in neuronal membranes, e.g. *Phyllobates terribilis*, which is protected from predation by secretions from glands in the skin that contain toxins that prevent the sodium channels from closing at the end of the action potential.

(b) The mechanism of synaptic inhibition was first studied in the neuromuscular junctions of lobsters and crayfish. In vertebrates, motor neurons are always excitatory, but many arthropod muscles are innervated by both inhibitory and excitatory motor neurons. Those in the lobster claw are particularly large and accessible. Crustaceans, like insects, have a haemocoel in which the blood at low pressure bathes the tissues, so the muscles and neurons survive well when exposed or when isolated from the body.

(c) Krait venom is used for studying the mechanism of transmission at vertebrate neuromuscular junctions. Kraits prey on other vertebrates and the major ingredient of their venom binds irreversibly with the receptor sites for acetylcholine, thereby blocking neuromuscular transmission and paralysing the prey.

Question 2 The correct answers are (d), (e), (g) and (h). (a) is wrong because cephalopods, particularly large, actively swimming species like squid, are more difficult to keep in captivity, and require more elaborate facilities than mammals. Mammalian behaviour is more diverse and complex than that of cephalopods so (b) is wrong. The mammalian nervous system is larger, relative to total body mass, than that of cephalopods so (c) is wrong. (f) is wrong because cephalopods have small neurons as well as large ones, although little is known about their properties because they are much more difficult to study. The giant neurons, about which a great deal is known, represent a small minority of the total number of neurons in the squid nervous system.

Question 3 (a) Data from young laboratory rodents suggested that the principal mechanism of adipose tissue expansion is enlargement of a population of adipocytes that increases in numbers by at most twofold. However, non-experimental, indirect measurements on a heterogeneous population of people do not fit well into this concept. The poor fit could be due to the data being inaccurate, or to the concept derived from rodents being inapplicable to humans. If a natural system of classification is used, taxonomic status indicates the degree to which organisms have evolved independently of each other. Increase in the numbers of adipocytes also makes a major contribution to adipose tissue expansion in other primates (i.e. monkeys), indicating that there may be a fundamental difference between rodents and primates in the cellular mechanism of adipose tissue expansion. Rodents are therefore less suitable as models of the human condition for studying this aspect of the physiology of adipose tissue.

(b) Female polar bears produce large quantities of lipid-rich milk while fasting, but the distribution of their adipose tissue is not significantly different from that of males. Therefore, even in a mammal that breeds while fasting, the energy demands of reproduction and lactation do not require sex differences in the distribution of adipose tissue, making such a theory less probable as an explanation of sex differences in the distribution of adipose tissue in humans.

Question 4 (b), (c) and (d) are true. (a) is false because some viruses proliferate in alternative hosts, and some animal parasites and bacteria are host-specific.

Question 5 All except (e) are false. (a) Although many prokaryotes and fungi are biochemically very versatile, many plants and some animals (e.g. certain spiders, snakes and insects) produce very powerful and specific toxins. However, because prokaryotes and fungi are easier to breed in captivity than large plants and animals, the drugs that they produce are often preferred for commercial exploitation. (b) Most snake and fish (e.g. puffer fish) venoms are toxic to many other vertebrates, including, as in the case of *Bungarus*, other species of the same family. Toxins produced by actinomycetes kill other bacteria. (c) Drugs such as antibiotics that are effective against pathogenic bacteria are produced by organisms of a wide range of taxonomic affinities, e.g. other bacteria, certain fungi, certain dipteran larvae (i.e. blowfly maggots). Conversely, closely related species (e.g. other fly larvae such as those of daddy longlegs and *Drosophila*) do not produce such substances. (d) There is such a wide variety of toxins, and many organisms are protected from predation by other means (e.g. by being mechanically tough or prickly (Figure 3.13), or by being inconspicuous or by running away) that it is impossible to predict exactly their behaviour, physiology or biochemical capacities. (f) This statement is often, but by no means invariably, true; e.g. quinine in bark limits proliferation of *Plasmodium*, the parasitic protistan that causes malaria in humans, although the tree and the protistan do not interact in the wild; antibiotics kill many species of bacteria that never interact naturally with the fungus or bacterium that produces them.

FURTHER READING

Alexander, R.McN. (1990) *Animals*, Cambridge University Press. Hardback ISBN 0521343917; paperback ISBN 052134865X.

Arms, K. and Camp, P. (1987) *Biology*, 3rd edn, Holt, Reinhart and Winston. ISBN 0030036445.

Barnes, R.D (1987) *Invertebrate Zoology*, 5th edn, Saunders College Publishing, Philadelphia, New York, London. ISBN 003008914X.

Barnes R.S.K., Calow, P. and Olive, P.J.W. (1988) *The Invertebrates: A New Synthesis*, Blackwells Scientific Publications, Oxford. ISBN 0632016388.

Langley, G. (1989) *Animal Experimentation: The Consensus Changes*, Collier Macmillan Publishing Co., New York, London. Hardback ISBN 0333453824; paperback ISBN 0333453832.

Muller, W.H. (1982) *Botany: A Functional Approach*, 4th edn, Collier Macmillan Publishing Co., New York, London. ISBN 0029794404.

Nester, E.W., Roberts, C.E., Lidstrom, M.E., Pearsall, N.N. and Nester, M.T. (1983) *Microbiology*, 3rd edn, Holt, Reinhart and Winston. ISBN 483370143X.

Pough, F.H., Heiser, J.B. and McFarland, W.N. (1989) *Vertebrate Life*, 3rd edn, Collier Macmillan Publishing Co., New York, London. ISBN 0023963603.

Romer, A.S. and Parsons, T.S. (1986) *The Vertebrate Body*, 6th edn, Holt, Reinhart and Winston. ISBN 0030584469.

Stace, C.A. (1989) *Plant Taxonomy and Biosystematics*, 2nd edn, E. Arnold. Hardback ISBN 0713129808; paperback ISBN 0713129557.

Wistreich, G. (1988) *Microbiology*, 5th edn, Collier Macmillan Publishing Co., New York, London. ISBN 0024289507.

ACKNOWLEDGEMENTS

The series *Biology: Form and Function* (for Open University Course S203) is based on and updates the material in Course S202. The present Course Team gratefully acknowledges the work of those involved in the previous Course who are not also listed as authors in this book, in particular: Ian Calvert, Lindsay Haddon, Sean Murphy and Jeff Thomas.

Grateful acknowledgement is made to the following sources for permission to reproduce material in this book:

FIGURES

Figure 1.6: Parsons, T. R. and Takamashi, M. (1973) *Biological Oceanographic Processes*, Pergamon Press Ltd; *Figures 1.7 and 1.8:* Hyman, L. H. (1940) *The Invertebrates*, McGraw-Hill Book Company; *Figure 1.24(a):* Meglitsch, P. A. (1967) *Invertebrate Zoology*, Oxford University Press; *Figure 2.9:* Romer, A. S. (1962) (3rd edn) *The Vertebrate Body*, Saunders College Publishing; *Figure 2.11:* Courtesy of Dr Peter Sheldon, Department of Geology, University of Wales, Cardiff; *Figure 2.12:* Sedgwick Museum, University of Cambridge, SMJ 37289–37290; *Figure 2.13:* Jensen, W. A. and Salisbury, F. B., *Botany: An Ecological Approach*, copyright © 1972 by Wadsworth Publishing Co., Inc., reprinted by permission of the publisher; *Figure 2.17:* Romer, A. S. (1966) (3rd edn) *Vertebrate Paleontology*, The University of Chicago Press; *Figure 3.1:* Courtesy of Mr M. A. King; *Figures 3.5 and 3.6a:* Courtesy of Dr Michael Stewart; *Figure 3.9:* Pond, C. M., Mattacks, C. A., Thompson, M. C. and Sadler, D. (1986) *Brit. J. Nutr.*, **56**: 29–48; *Figure 3.10:* Sjöström, L. and Björntorp, P. (1974) 'Body composition and adipose tissue cellularity in human obesity', *Acta. Med. Scand.*, **195**, Almqvist & Wiksell International; *Figure 3.11:* Pond, C. M. and Mattacks, C. A. (1987) *Folia Primatologica*, **48**: 164–185; *Figure 3.12:* Mr Marc Cattet (1988) 'Aspects of physical condition in black bears and polar bears', MSc. thesis, University of Alberta.

PLATES

Plates 1, 2(b), 3(a)–3(d): Courtesy of Dr H. Canter-Lund; *Plates 4, 5(a), 5(b), 6(a), 6(b), 7(a)–7(e), 11(b):* Courtesy of Dr J. E. Bebbington; *Plates 8(b), 11(a)–11(g), 11(i), 11(j), 11(m), 11(n), 12, 13(b), 13(d), 14(a), 15(c), 15(e), 15(h), 15(i), 16(c), 16(e), 16(f), 16(i)–16(k), 17(a), 17(c):* Courtesy of Professor D. H. Janzen; *Plates 10(c)–10(e):* Courtesy of Dr Sinclair Stammers; *Plates 10(b) and 10(f):* Courtesy of Dr A. Lister; *Plates 3(e), 6(c), 8(b), 9, 10(a), 11(h), 13(a), 13(c), 14(b), 14(c), 15(a), 15(f), 15(g), 16(a), 16(b), 16(d), 16(g), 16(h), 17(b), 17(d):* Courtesy of Dr C. M. Pond.

INDEX

Note Entries in **bold** are key terms, page numbers in *italics* refer to figures, tables and plates.

192